Contents

Editorial 2
Caroline Sweetman

Liberalisation, gender, and the land question in sub-Saharan Africa 9
Kaori Izumi

Does land ownership make a difference?
Women's roles in agriculture in Kerala, India 19
Shoba Arun

Rural development in Brazil: Are we practising feminism or gender? 28
Cecilia Sardenberg, Ana Alice Costa, and Elizete Passos

Women farmers and economic change in northern Ghana 39
Rachel Naylor

'Lazy men', time-use, and rural development in Zambia 49
Ann Whitehead

Integrating gender needs into drinking-water projects in Nepal 62
Shibesh Chandra Regmi and Ben Fawcett

Structural adjustment, women, and agriculture in Cameroon 73
Charles Fonchingong

Interview with Penny Fowler and Koos Neefjes of Oxfam GB:
Are genetically modified foods a new development? 80

Resources: 86
compiled by Erin Murphy Graham
   Books and papers 86
   Organisations 88
   Web resources 89
   E-mail lists 90

Index to Volume 7 91

Editorial

As we enter the new century, there is an increasing trend towards the 'feminisation of agriculture' (FAO 1999i). The role of women in food production is expanding: in south-east Asia, women currently provide up to 90 per cent of labour for rice cultivation, while in sub-Saharan Africa, women produce up to 80 per cent of basic foodstuffs for household consumption and sale (FAO 1999ii). One key element in this process is the fact that rural livelihoods are changing, as a result of economic crisis and growing pressure on scant natural resources. For example, male migration has led to a 21.8 per cent drop in the rural male population in Malawi between 1970 and 1990, while the female population declined by only 5.4 per cent (ibid.). The role of women is also expanding to compensate for 'missing men', lost in armed conflict and through disease, including AIDS. In sub-Saharan Africa, AIDS is devastating agricultural production, and there is an increasing dependence on households headed by females, children, and the elderly. The 1999 harvest in Zimbabwe saw a 61 per cent decrease in maize output (*Mail and Guardian*, 16 August 1999). Agribusiness is also being 'feminised': in countries like Chile, where overall economic statistics reflect the achievements of agribusiness in bringing about an economic 'miracle' (Barrientos et al. 1999, 1), women workers have become central to the production of fruit for export, providing 'flexible female labour ... with a clear pattern of gender segregation' (ibid., 9).

Articles in this collection assert that women's contribution to global agricultural production for food and for profit continues to be largely unacknowledged and undervalued, and that their ability to farm is constrained, because the resources they need are often controlled by others. Women in many different contexts continue to have their rights denied to independent control of land, agricultural inputs, credit, and other essential resources. Their access to training, education, and extension services, and to gaining leadership of rural organisations are impeded by assumptions on the part of national governments, community leaders, and development policy-makers that farmers are male, because 'men are the providers'. New technologies which are available to male farmers may not be available to women, while women's own knowledge of crops and husbandry is either underestimated, or appropriated by private companies which can profit from it. New agricultural policies are needed, founded on a gender analysis of the process through which agriculture is becoming feminised, and a commitment to gender equality.

Agriculture, production, and gender relations

The high productivity and low visibility of women in agriculture first received worldwide attention 30 years ago with the publication of Ester Boserup's book *Woman's Role in Economic Development* in 1970. The UN International Decade for Women (1976-85) subsequently focused on women's role in production.

The main focus of Boserup's book is the impact on international development of a failure to recognise the extent of women's responsibilities, and to support women in this. During International Women's Decade, feminist researchers shifted the emphasis to gender equity, to focus on the ways in which changing conditions of production, and shifts in the division of labour, are linked to changes in women's status — for better or worse.

These studies drew the attention of policy-makers to the enormous workload of women across the world, and contrasted this to women's lack of control over the land and property they used in production, and their lack of a say in how the products of their work were used. These analyses questioned racist stereotypes of rural societies as backward, by pointing out women's similar experiences in the industrial, 'modern' settings of Europe and North America. Cross-cultural comparisons showed that women's role in production was under-valued everywhere: a shift away from farming to manufacturing and service industries did not necessarily end inequality between the sexes. These studies also exposed the fact that women were — and still are — burdened with almost all the domestic work and child-rearing throughout the world, and that this is linked to the lesser value ascribed to women's and men's work.

Rural development interventions focus primarily on promoting efficiency in the agricultural sector, rather than promoting equality between the sexes. However, all the articles in this issue assert that a focus on gender equity is essential, even if the aim is only to increase efficiency. It is clear that (in societies where agriculture is the sole or the major source of household livelihoods) modes of production are related to the division of labour within the household, and in particular to marriage and family forms. The implications of this were clearly recognised by Ester Boserup: 'Economic and social development unavoidably entails the disintegration of the division of labour among the two sexes traditionally established in the village' (Boserup 1989, 5). Feminist analysis also confirms the links between underlying power relations between the sexes, which define, and are defined by, the gender division of labour. For example, norms of female submission and fidelity within marriage are an economic, as well as a social, issue: control of women's bodies is essential if men are to be certain of the paternity of the children who will inherit their land and property. This control in turn shapes women's participation in production, since their mobility outside the household may be restricted and policed. Forms of marriage — monogamy, polygamy, and polyandry — determine the size and nature of the household labour force, and the resources available to the household are determined by different systems of land and property inheritance, forms of marriage, and norms of access and control.

Women's land rights: access, control, and ownership

Independent land rights, which enable women to decide on the use of land and keep the proceeds from such use, are still a dream for women in many countries, despite their increasingly central role in agriculture. Women's relationship with land is determined by customs and laws of inheritance and marriage. If a woman does not inherit her father's property, but is expected instead to marry and move to her husband's land, she only has access to the land of her natal and marital homes. In some contexts, women do keep the proceeds from the crops they grow and sell on their husband's property; but without formal ownership of land, they are barred from using it as collateral for loans or credit, selling it if they have to raise money, or bequeathing it to daughters or others.

The need for women to secure full and independent land rights has been argued on the grounds of welfare, efficiency, and gender equality (Agarwal 1994). On welfare grounds, landlessness has been linked in many studies to poverty in South Asia. In comparison, landlessness in sub-Saharan Africa has been comparatively rare until now. However, the AIDS epidemic is causing distress sales of land in Zimbabwe, because many families cannot make a living from the land due to a lack of labour, and need to pay for medical care. It is likely that a new group of landless will arise from this problem (*Mail and Guardian*, 16 August 1999). As can be seen from this example, land is not only valuable for its use in agriculture, but is also a marketable commodity which provides security in times of crisis. In terms of efficiency, the 'women and environment' approach to development (WED) adopted by some development organisations argues that women are more likely to use land productively and sustainably. Empowerment approaches stress the fact that land ownership is not only an economic issue, but is closely correlated with social and political power.

However, ownership of land does not guarantee control over land or power in the home. In her article, Shoba Arun focuses on Kerala in southern India, which is well-known to gender and development researchers and workers for its relatively high level of social development, against a background of poverty. A feature of Kerala's *nair* communities is women's ownership of land, but control over it is either joint or determined by the natal family (Agarwal 1994). In Kerala, husbands and wives are also increasingly moving away from their land to gain male employment, and land is passing out of women's ownership. In other households, male migration has increased, and the impact of this is similar for women in both matrilineal and patrilineal households: both groups of women supervise farming, with little infrastructural support, and juggle this with their household work.

Today, women in many areas are unable to obtain rights to land by recourse to the law. While rights to equality are enshrined in many constitutions, laws do not always match this commitment, and where laws do exist they may not be accessible for women living in poverty. In India and many parts of sub-Saharan Africa, customs on land use and ownership were codified during the colonial era. Traditional customs had been relatively flexible and open to interpretation, but colonial officials who consulted male elders recorded their versions as 'customary law'. Currently, the situation is worsening for women in some countries; advances made in civil law are undermined by forces who wish to consolidate their power through asserting that 'custom' does not give women equal rights. Earlier this year, the international women's movement was horrified by the decision of the Supreme Court of Zimbabwe that Venia Magaya could not inherit her father's estate, citing customary law as a reason for treating women as 'junior males' (*Southern Africa Chronicle*, 31 May 1999). The political and economic background to Zimbabwe's ruling is discussed in Kaori Izumi's article here, and compared to the case of Tanzania.

Izumi also explores the different aims of various phases of land reform policy. During the first years after the colonial era, many countries emphasised redistribution of land to people living in poverty.

However, during the economic adjustment drives of the 1980s, with their political ideologies of deregulation and individual competition, women and other marginalised groups have lost out. Debates on land and gender equity tend to be limited to discussing the pros and cons of civil law and individual rights versus collective ownership and customary law. In her accessible discussion of land theories, Izumi shows how these reflect the historical context and current political and economic ideologies.

Agriculture, livelihoods, and economic adjustment

Over the past decade, world trade has expanded, but while many countries have increased their agricultural exports, others (including most in sub-Saharan Africa) have not been able to take advantage of the opportunities of global trade. The 48 least developed countries, home to 10 per cent of the world's population, have seen their share of world exports decline to 0.4 per cent over the past two decades. In contrast, the United States and the European Union contain roughly the same number of people, yet account for 50 per cent of world exports (World Bank 1998). Poor farmers not only are unable to gain access to global export markets for cash crops, but also face other threats to their livelihoods, health, and food security. The activities of multinational corporations, promoting patented technologies developed through genetic modification, pose one such threat.

The impact of globalisation on farmers varies according to their context and social identity. Two articles in this collection discuss how structural adjustment has affected African agriculture, and women's lives. Rachel Naylor discusses Ghana, where path-breaking feminist research on patterns of agricultural production and household budgeting was carried out by Ann Whitehead almost two decades ago. Naylor points out that the language used in discussions of women's vulnerability, of the impact of economic reform, and of the 'rural poor', renders rural dwellers passive, and obscures the fact that people adjust to meet challenges posed to them. Yet while men and women are taking advantage of new opportunities, the evidence shows that in Ghana, as elsewhere, women are also shouldering added burdens. In his article, Charles Fonchingong charts how members of women's self-help groups in Cameroon perceive life during the structural adjustment programmes of the past decade. In this case study, women from rural areas describe how they see the formerly clear divisions between male and female systems of agriculture blurring into one, as household members struggle to overcome the threat of poverty brought about by structural adjustment.

Several articles in this issue discuss agricultural production as one component of increasingly diverse livelihoods in rural and (to a lesser extent) urban areas. Migration — already a key component of rural livelihoods for many — has become increasingly significant in this era of environmental degradation, pressure on scant resources, and structural adjustment. Many in rural areas have no sources of cash, and subsistence cannot be guaranteed, so they have to maintain living standards through cash employment. Migration is a gender-specific issue: depending on the nature and conditions of the work available, either women or men will travel to find it. Age, and the stage of the family life-cycle, also determine who goes and who stays. In central and south America and east Asia, women are often the migrants; in south Asia, the picture is different. In their article focusing on an arid area of Brazil, Cecilia Sardenberg, Ana Alice Costa, and Elizete Passos focus on a major project which aims to alleviate poverty among rural dwellers. They point out that 'loss of labour through migration can effectively double women's workload: in certain communities, nearly 80 per cent of the households and the care of the land are under women's responsibility for the greatest part of the year' (Sardenberg et al., this issue).

The second green revolution

From the mid-1960s onwards, rural development research and policy concentrated on the need to 'modernise' the agricultural sector, promoting the cultivation of cash crops on large-scale land-holdings. The first green revolution saw the advent of high-tech 'solutions' to food insecurity in Asia and Africa, including chemical fertilisers and pesticides, mechanised irrigation, and new high-yield crop varieties developed in the laboratories of North America and Europe.

At present, consumers and environmental activists are challenging the right of private companies to shape the second green revolution, with an eye to profit rather than the goal of human development through global, sustainable, food security. While international development agencies and national governments controlled the first green revolution, the second green revolution is being shaped by the will of multi-national corporations, and by specific governments which are pursuing the new technologies (the US is currently exporting $50 billion of agricultural products a year and planting transgenic varieties for 25-45 per cent of its major crops, according to UNDP 1999 figures[1]). There is currently an international outcry, caused by a concern for public health, for the right of states to ensure food security of their populations, and by the threat of environmental hazard and the appropriation of communal resources by private ownership. In an interview with Koos Neefjes and Penny Fowler here, these issues are explored, and the connections between this debate and gender issues teased out.

Vandana Shiva, who is perhaps the best-known feminist environmental activist today, highlights the links between the ideologies which determine the course of global development, the damage which development interventions have wreaked on the environment, and women's well-being and status.

For gender and development researchers and workers, ecofeminist approaches which believe in the close connection between women and nature (Agarwal 1992) may seem naïve in their failure to analyse the way in which women are divided as a group by other aspects of their identity, and in their conflation of 'women's perspectives and actions' with other 'alternative' visions of agriculture and world trade (Shiva 1996, 26). Data on the impact of the first green revolution of the 1960s and 1970s show that 'the major beneficiaries are those who were already relatively well off … but there is no simple opposition between men and women' (White 1992, 46). Key elements of Shiva's recent work on globalisation agree with the views of activists from anti-poverty and environmental organisations: for example, that states should have the right to feed their people without competing with global players in a so-called free, but unfair, market, and that consumers should 'think globally, act locally'. While it is true that 'in the North and the South, women have been in the forefront [of the struggle] against industrial farming methods which destroy livelihoods and ecosystems' (Shiva 1996, 25), women involved in political protest on these issues have more in common than their sex.

Analytical tools and women's workload

Two articles in this issue shed light on important methodological issues for development policy-makers and practitioners. A number of development organisations and individuals have developed analytical tools to assist the process of integrating gender issues into planning and implementation. However, a major risk in using such tools is that they can be applied mechanistically, without commitment to challenging injustice (Smyth 1999).

Another risk is that inaccurate data are produced and development interventions informed by them. Studies of women's and men's time-use formed a vital element of the pioneering research into the role of women in agriculture of the 1970s and 1980s. Data of this type has been used to raise awareness among community groups and development practitioners of the unequal workloads of women and men, and of women's multiple roles in productive and reproductive work. In an article which reviews an influential paper on time-use in Zambia, Ann Whitehead points out the need to understand the role of agriculture in rural livelihoods, and the gendered nature of the external employment market, before making assumptions about women's and men's roles in production. Whitehead argues that in the absence of such a detailed understanding of context, skewed figures can result in a lack of understanding of rural livelihoods and in inaccurate stereotypes of African men as lazy.

Women's central role in contributing to rural (and urban!) livelihoods through domestic work is rarely shared by men. The need to alleviate the time-consuming drudgery of water and fuel collection in developing countries is widely accepted as essential if welfare goals are to be met, and rural development initiatives to be rendered sustainable. Shibesh Regmi and Ben Fawcett's article, focusing on water provision in Nepal, criticises the limited understanding of 'gender' or 'women's' issues on the part of development practitioners involved in technical aspects of rural development. This article offers useful insights for other technical specialists involved in aspects of rural development, who commonly consider 'strategic'(Moser 1989) — or feminist — issues of gender power relations to be outside their remit. The language of many rural development initiatives speaks of women's 'practical needs' and 'women's participation', but the links between practical needs and strategic issues — including control over essential resources — are lost. Regmi and Fawcett document how a water project can fail in the absence of an understanding of how gender relations affect a community's chances of attaining sustainable development.

To close, there is one final stereotype persisting in rural development to be challenged, which is extremely influential in determining what kind of work with women in rural areas is appropriate for NGOs and government bodies to attempt. This stereotype is one of exhausted, victimised, and uneducated rural women, who are victims of back-ward, traditional forces in their households and at community level, and who are unaware of the obstacles they face. Issues including domestic violence may be tip-toed around by development workers who are anxious to focus on meeting 'basic needs' and chary of broaching sensitive issues which they cannot address through tangible work. Staff promoting social development are often located in urban areas, in the belief that gender inequality can be challenged better in this setting, and lack the understanding to challenge rural reality.

Articles in this collection make suggestions on ways to 'work on gender' as an essential component of all development initiatives, and highlight the impact on the food security and wellbeing of women, men, and children in rural areas if women's interests are disregarded. There is compelling evidence here that economic, legal, and social aspects of women's poverty, and especially their rights to land, must be challenged simultaneously.

Note

1 UNDP Human Development Report 1999, 72.

References

Agarwal B (1992) 'The Gender and Environment Debate: Lessons from India', in *Feminist Studies*, 18:1.

Agarwal B (1994) *A Field of One's Own: Gender and Land Rights in South Asia*, Cambridge University Press.

Barrientos S, Bee A, Matear A and Vogel I (1999) *Women and Agribusiness: Working miracles in the Chilean fruit export sector*, Macmillan, London.

Boserup E (1989) *Women's Role in Economic Development*, Earthscan, London.

FAO (1999i) http://www.fao.org/Gender/en/agrib2-e.htm

FAO (1999ii) http://www.fao.org/Gender/en/agrib4-e.htm

Mail and Guardian, South Africa, http://www.mg.co.za

Moser C (1989) 'Gender planning in the Third World: Meeting women's practical and strategic gender needs', in *World Development* 17(11).

Shiva V (1996) *Caliber of Destruction: Globalisation, food security and women's livelihoods*, Isis International, Philippines.

Smyth I (1999) Introduction in *A Guide to Gender-Analysis Tools and Frameworks*, Oxfam GB, Oxford.

UNDP (1999) *Human Development Report*, UN, New York.

White S C (1992) *Arguing with the Crocodile: Gender and class in Bangladesh*, Zed Books, London.

World Bank (1998) *World Development Indicators*, World Bank, Washington DC.

Liberalisation, gender, and the land question in sub-Saharan Africa

Kaori Izumi

This paper focuses on land reform initiatives undertaken in a number of African countries since the late 1980s. Current theories of land and debates on gender issues fail to explain the complex processes through which women's access and rights to land have been affected, contested, and negotiated during socio-economic and political restructuring. Drawing on the case studies of Tanzania and Zimbabwe, this paper is a call for policy-makers, researchers, and activists to return to these neglected issues.

In the process of social, economic, and political restructuring that most African countries have undergone in the past two decades, land has been one of the most contested issues. Privatisation of land has become the major objective of land reform in a number of African countries — including Tanzania, Mozambique, Malawi, Zambia, Botswana and Namibia — where economic adjustment policies imposed by the International Monetary Fund and World Bank aim to allow market forces to determine the efficient allocation and use of land. This shift in the direction of land policy has also affected countries like Zimbabwe and South Africa, where redistribution of land to the black majority was the original rationale of land reform.

Land policy formulation has become an arena of conflict for a number of interest groups[1]: at the local level, land-related conflicts have arisen and intensified for reasons including pressure on land-use, and the investment potential of particular areas. Political parties have used the question of the direction, the pace, and the way in which land reform is instituted, as a means of acquiring support in new multi-party systems. Political conflict over land has emerged between ethnic groups, as well as between national and local state institutions.

In this article, through a review of the cases of Tanzania and Zimbabwe, and a discussion of both mainstream theory[2] and gender analysis of land issues, I will discuss how economic and political liberalisation have affected women's access and rights to land. Two concerns are of particular importance: first, the question whether market-driven land reform can be compatible with a goal of equal distribution of land among vulnerable groups, including poor people, the landless, and women. Debates on land reform in the context of economic liberalisation have tended to omit gender-related issues. This omission is also reflected in the process of actual land reforms in the 1990s: because the logic of the market is to promote maximum efficiency through competition, it is indifferent to issues of equity. Gender issues have been largely sidelined and compromised, and control of land has been retained by existing powerful social groups.

The other, related, concern is to track why particular groups are vulnerable as regards land. During the period of economic liberalisation, women's rights to land have been affected by state policies, market forces, and traditional social structures, in different ways and to different extents. It is not only the kind of institutions which affect women's access and rights to land, but the power relations which exist in different institutions — national and local state structures, the market, the community or village, and the family and household (Agarwal 1994). These power relations are determined by cross-cutting factors such as gender, class, race and ethnicity.

Changes to land tenure during liberalisation

A principle of African indigenous land tenure is to protect the access to land of members of a family, and members of a community. In patrilineal societies (where inheritance passes through the male line), women's rights to land are usually determined by their relationship to men. However, the forms of indigenous land tenure are highly diverse; in some communities, women have relatively secure access to land (Lastarria-Cornhiel 1997: 1320). Indigenous land tenure has changed over time due to various factors (including population pressure, commercialisation of agriculture, increasing land sales, opening of investment opportunities in land, urban-expansion, AIDS, and land-grabbing). In areas of acute land shortage, where the value of land has increased, women tended to lose their customary rights (ibid., 1325).

Economic liberalisation seems to have further accelerated such processes, both in pace and in scale. In many contexts, liberalisation has opened new opportunities for investment in land, leading to increased and intensified contests over land. As land gains value as a commodity for investment, land-grabbing by political elites, appropriation of village land by the state, and allocation of land to investors — both nationally and locally — have become common phenomena (Moyo 1998, Kanyinga 1998). On the other hand, the poor are selling land as a desperate act for survival as poverty deepens. Retrenchment, unemployment, and declining real wages due to high inflation are forcing urban workers to search for rural land in order to produce food to supplement their low income.

Theories on land in Africa: A gender analysis

How, and to what extent, can existing theoretical insights on land issues offer a useful analytical framework for understanding gender issues relating to land? In this section, I will briefly examine the gender-blind 'mainstream' theoretical debates on land, and the gender-sensitive studies of land issues.[3]

Evolution

The desirability of individual land ownership, titling, and registration has for a long time dominated the debate on land tenure in Africa. Efficient resource allocation by individual agents is understood to require both tradable land property, and the development of a market for land. Conventional theories of individual land ownership, titling, and registration anticipate the state playing an active role in replacing indigenous land tenure systems with individualised land tenure (Falloux 1987). Recently, this has been challenged by the idea of the 'evolutionary position' of African land tenure. This idea sees African land tenure systems as flexible and adaptable to changing conditions, and its advocates argue that they will evolve naturally towards individualised land tenure. Thus, drastic state intervention is considered unnecessary (Bruce and Migot-Adholla 1994, Platteau 1996)[4]. The evolutionary position has had increasing influence on World Bank land policy in Africa since the late 1980s (World Bank 1989, 1992).

The evolutionary theory of African land tenure is built on a number of assumptions: that individualised tenure would lead to more efficient use of land and higher security; that local communities in Africa are homogeneous; and that the social institutions which administer indigenous land tenure systems are adaptable and flexible. But it fails, first, to recognise that power and social relationships between certain categories of people determine their interests in, and access to, land. Second, it ignores the fact that change does not take place as a spontaneous evolutionary process towards greater efficiency and security for all, but that such processes are in the interests of powerful stakeholders. For rural women, an 'evolutionary' process of land tenure has often meant more insecurity in access and rights to land, as a consequence of commercialisation of agriculture and individual land titling which discriminates against them (Lastarria-Cornhiel 1997).

Negotiation

Another theoretical position holds that land rights and access depends on people's capability to negotiate, manipulate rules and norms, and to 'straddle' different institutions (Berry 1993 and 1994, Moore 1999). Because multiple institutions determine people's relationship with land, they are potentially able to find multiple opportunities for political manoeuvring, 'straddling' different institutions: informal and formal, political and legal, traditional and modern.

In the negotiation approach, access and rights to land are discussed in relation to an individual's capability, neglecting structural constraints on that individual. People's access to economic and political power, and their capabilities for negotiation, differ by gender, class, race, and ethnicity. Moreover, power relations shape the terms in which negotiation takes place, and the forms how individuals actually negotiate in different arenas.

Democratisation

Democratisation theory asserts that the land question cannot be resolved either by depending on indigenous land tenure to 'evolve' by itself, or through individual negotiation with institutions. Rather, it highlights the contradictory power relations between various actors and social groups, including the central and local state, various types of private investors, small-holders, pastoralists, ethnic groups. Theorists analyse the way in which the land rights of small-holders have been undermined through state interventions and in the process of economic and political liberalisation (Amanor 1997i and ii, Kanyinga 1997). In this position, the solution is seen to be collective political action by small-holders towards democratisation and decentralisation.

Because of its primary concern about class, the democratisation approach fails to consider the role of patriarchal power in influencing women's participation and representation in political activity. Other questions which remain unanswered include how successful women have been in influencing policies and decisions on land; what structural constraints they face in organising and lobbying for land access and rights; how women's interests in land differ according to class, ethnicity, and race; how their opinions and interests are integrated in wider struggles for democratisation and decentralisation; and whether democratisation and decentralisation in different contexts revive and reinforce traditional authorities, or lead to new democratic institutions, which address gender issues.

Gender perspectives on land issues in Africa

The limited debates on women's land access and rights in Africa have so far focused on the implications for women of two systems of land tenure: individual and indigenous (or customary, or communal). In other words, one could characterise this debate as concerned with the market versus tradition. Some see indigenous land tenure as placing constraints on women; in contrast, the process of 'individualisation' is assumed to provide women with equal opportunities in access to land (Mac Auslan 1996).

Others are sceptical about the opportunities offered by the free-market solution. Their argument is that the market is not gender-neutral, and moreover excludes poor women by discriminating against them because of their lesser power and resources (Meisen-Dick et al. 1997, Lastarria-Cornhiel 1997). This view does not necessarily idealise indigenous land tenure, but it does point out that, in the process of privatisation, women are losing some rights which they held under indigenous tenure.

The cases of Tanzania and Zimbabwe

Gender studies on land have focused mainly on the opportunities and constraints afforded to women of two forms of land tenure: individual ownership, and indigenous land tenure. In Tanzania, gender issues did not command sufficient attention from policy-makers in planning land reform. In contrast, in Zimbabwe, economic and political forces have gradually sidelined gender issues of land, while the formal policies and laws relating to gender equality are failing to enforce women's formal rights in practice.

Tanzania

Between 1973 and 1976, under the policy of *ujamaa* ('villagisation') 13 million people were forced to settle in 8,000 villages. As a response to the changing conditions brought about by structural adjustment, which was adopted in 1986, institutional reforms for land were initiated at the end of the 1980s[5].

Three different documents have been produced in the process of land reform in Tanzania: a 'Report of the Presidential Commission of Inquiry into Land Matters' (1994), a National Land Policy (1995), and a Draft Bill for the Land Act (1999). In the first of these, the Land Commission (appointed in 1991) investigated land matters and made policy recommendations. The primary concerns of the Commission were security of land rights among villagers, and democratic management of village land by village assemblies. In terms of gender concerns over land, however, the Land Commission was reluctant to make any radical change except a provision of joint ownership of land between spouses (URT 1994).

The National Land Policy of 1995 completely sidelined the Commission's recommendations. Instead of democratising and decentralising land management, the National Land Policy is seen by many as centralising and strengthening the power of the state to control land (URT 1995). The principle of joint ownership between spouses, which was agreed upon during a national workshop on land in 1995, was deleted from the final policy.

In 1996, a Bill for the Land Act was drafted by a British consultant, sponsored by the Overseas Development Administration (ODA) of the British government. In a speech given in November 1996,[6] he stressed the need for an 'efficient and equitable land market' (Mac Auslan 1996, 4), for the 'involvement of private sector in legal reforms for land' (ibid., 7), and for people to have 'freedom to enter and undertake transaction in the land market' (ibid., 15). Here, the private sector was used as a synonym for civil society (ibid., 6-8). The Draft Bill stressed that the Bill is concerned with equality of opportunity; thus it was not designed to provide women with preferential treatment, but to prevent women being given less favourable treatment than men (ibid., 10). Women's organisations and female lawyers supported the principle of the Bill as a step towards equal opportunity for women (Rwanbangira 1997, 1; TAWLA 1997).

However, these supporters have been criticised for failing to argue that the market discriminates against individuals and social groups according to their resources and capabilities. The women's organisations supporting the Draft Bill were predominantly urban-based, educated, and middle-class, so the principles of individual land ownership and the development of a market-based system were in line with their interests, as distinct from those of rural women (Manji 1998).

Only minor ad-hoc amendments to the Bill were requested, and the chance was missed to address the fundamental shortcomings of the Bill from the perspective of women living in poverty (TAWLA 1997). A coalition of NGOs against the Bill lobbied against the Bill going through Parliament without public debates, but the Bill was finally passed in February 1999.[7]

In Tanzania, a period of economic and political change over the past decade has contributed to intensifying demand and conflicts over land. At the local level, demands and conflicts over land have intensified, not only between the private and state sectors on one hand and peasants on the other, but also within local communities, families, and households. Conflicts over land rights have been most heated in areas which offer potential for investment, including peri-urban areas, commercial agricultural areas, coastal areas, and mining areas.

Although systematic data is not available, my own research during 1996 in one village in Morogoro showed that a village leader expropriated collective land belonging to women's groups and sold it to outsiders (Izumi 1998i). In Turiani, Morogoro region, where sugar-cane production has expanded because of measures introduced to promote the cultivation of cash-crops under structural adjustment, cases where husbands are selling land without their wives' consent are increasing (ibid.)[8].

In the areas where conflicts over land are intensifying, and where the market discriminates against women and men living in poverty, women are finding that customary land tenure, which formerly offered women some means of protection, is eroding rapidly. Widows, who were previously allowed to stay on the land after the death of their husbands, are now subject to dispossession. In Kagera region and in other areas where the HIV/AIDS epidemic is widespread, the male relatives of deceased husbands systematically dispossessed AIDS widows[9]. In cases where widows are not infected by AIDS themselves, they may be blamed for the death of their husband, and this blame is used against them as an excuse for dispossessing them. Divorced women, who were previously provided with a piece of land by their father, are losing such access to land in their natal community.

While it is clear that the indigenous land tenure systems now offer women no protection, the market is unlikely to offer a better alternative means of ensuring land rights for women living in poverty, who are unable to compete on an equal footing due to class- and gender-based discrimination.

Zimbabwe

During the colonial era in Zimbabwe, a racially-based skewed distribution of land excluded black Africans from access to land, so the struggle for independence from colonial rule focused on recovering the land that had been taken from the black population.[10] However, in the Lancaster House Constitution of 1979 it was stated that the state should purchase land for redistribution based on 'willing buyer – willing seller' principles. This principle informed the land-reform initiatives of the

1980s; it was a compromise, attempting to balance the interests of white large-scale commercial farmers, black Zimbabweans, and the demands of the new democracy.

During the 1980s, a number of new laws were passed promoting gender equality. The Legal Age of Majority Act of 1982 declared women (previously legal minors) to achieve legal majority at the age of 18 (GOZ 1982). In 1985, the Constitution of Zimbabwe was amended: Clause 4 stated that questions of citizenship were universal, regardless of gender and marital status (GOZ 1985). A State Party Report 1995 on implementation of the International Convention on Economic, Social and Cultural Rights confirmed that the government would ensure the equal rights of men and women (UNHCHR 1995).

Actual redistribution was only conducted on a limited scale during the first decade after independence. Zimbabwe adopted its Economic Structural Adjustment Programme (ESAP) in 1990, and the initial land-reform policy was gradually replaced by market-oriented land-reform. Land has gained in value as an investment asset; in 1992, the Land Acquisition Act provided compulsory acquisition of land at a government-controlled price for redistribution. The Act was first applied on a big scale in 1997, when 1,471 farms — 30 per cent of the total large-scale commercial farms — were designated.[11] Instead of waiting for the government to designate their land, large-scale commercial farmers have started to sell land privately.

ESAP has led to a focus on new, export-oriented land-uses such as wildlife management, horticultural cropping, livestock exports such as ostrich production, and tourism (Moyo 1998). The poor and landless have been largely excluded from the beneficiaries of land reform, and there has been a shift in debates on the direction and rationale for land redistribution from discussions of giving land to the black poor and landless on grounds of equity, to a focus on allocating of land to 'capable black farmers' (O'Flaherty 1998, 552; Moyo 1995i).

In resettlement areas, households predominantly consist of nuclear families who have relocated from their original villages. Resettlement land has been allocated to individual households as a unit, ostensibly because areas of land for redistribution are limited, but also because of male bias among land administrators (Jacobs 1996). Land was usually registered under the husband's name (Gaidzanwa 1995). Single women have been mostly excluded from land allocation, and women lose access to resettlement land upon divorce (Jacobs 1998, 279). 1997 data shows that about 75 per cent of the registered land owners are male, about 20 per cent of the farms were jointly owned, less than 5 per cent were owned by women, and below 4 per cent of land were owned by black women (Moyo 1997/8:31). However, in some respects, resettlement appears to have had a beneficial impact on women: in one study of resettlement areas in 1984, many wives reported an increase in the amount of land husbands allocated to them to cultivate, and a rise both in family and personal incomes (Jacobs 1998). Many women considered the fact that resettlement meant moving away from the husband's extended family to be positive, as it loosens the hold of their in-laws over them.

Frustrated by the slow pace of land reform, in some areas communities have started to occupy state-controlled land and resettlement areas illegally. Such actions have often been supported by local politicians and traditional leaders (Alexander 1994, Moyo 1998:18). In these processes, traditional leaders have been regaining their political power and legitimacy in communities.

As yet there is little documentation into how the current revival of traditional power has influenced Zimbabwean women's access to and rights to land within their community or village and the family/household, but some see this as having led the state to compromise on upholding women's land rights, in order to acquire the political support of local leaders (Moyo 1995ii; Gaidzanwa 1995, 8). In early 1999, the Supreme Court decided against the claim of Venia Magaya, the eldest child of her father's senior wife, who had been appointed heir to her father's estate in Harare by a community court. The court argued that under Zimbabwean customary law, which has coexisted with civil law since the colonial period, women are juveniles, and only men can inherit from a father (Supreme Court of Zimbabwe 1999). This ruling potentially undermines the principle of equality enshrined in the civil law, with grave consequences for all future rulings on women's rights.

A gender analysis of the land question in Zimbabwe shows the inability of formal law to ensure women's rights to land when such laws are not socially legitimate and enforceable. Empirical evidence from communal and resettlement areas shows that the relationship between traditional and modern forms of institution and women's access and right to land is complex and ambivalent in nature.

Conclusion

Many African societies have experienced substantial changes in their formal and informal land tenure systems, as part of a wider process of socio-economic and political changes. Land policy has become increasingly concerned with accommodating the free market, leading to a shift in focus from issues of poverty-alleviation, equity, and livelihoods to economic efficiency and investment. In this context, gender analysis of debates over land is one of the most neglected issues in research and policy debates. The mainstream theories regarding land issues in Africa are gender-blind, and this is reflected in the formulation of land policy and land law, which ignore the particular interests and needs of women.

Since the 1980s, both Tanzania and Zimbabwe have undergone a radical transition in terms of institutional land reforms. Their original objectives for land reform contrasted clearly, but this contrast has become less evident. Traditional institutions governing land tenure and use have been affected and transformed in different ways in these two countries, according to their past and present political and economic context.

Women's access and rights to land is shaped by gender-determined power relations, which exist across a range of institutions. The state, market forces, and tradition may interact, contradict, and cooperate in order to protect and strengthen existing power structures, which constrain women's secure access and rights to land. In Tanzania, traditional institutions have been transformed and undermined to a large extent in the areas affected by villagisation between 1973 and 1976. In Zimbabwe, a relatively strong and legitimate system of local government has coexisted with the central state, and has recently regained power as an alternative focus of political power.

Analyses of gender and land needs, therefore, need to go beyond the current common focus of weighing up the pros and cons of individual or indigenous land tenure. Further research is urgently required into the ways in which land access and rights among women have been affected, negotiated, and contested by — and within — the institutions of the state, the market, and the social institutions of the community, the family, and the household.

Several levels of analysis are necessary: two particular areas for attention are the gaps between statutory institutional reforms at national level, informal institutional changes, and actual practice at local level. This is because formal and legal rights to land do not necessarily provide women secure rights in reality, if such rights are not made socially legitimate and enforceable. At present, women in Zimbabwe have legal rights to land, but in many context they are without secure access. Institutions that govern women's relationship with land cannot be seen simply as a set of rules, norms, policies, and laws: it is the social legitimacy of these which constitutes institution. Because of this, women's access and rights to land is indeed a question of social change.

Kaori Izumi is a Visiting Assistant Professor at the Department of International Development Studies, Roskilde University, P.O. Box 260 DK-4000 Roskilde, Denmark.
E-mail: kaori@ruc.dk; tel: +45 (46) 742 322; fax: +45 (46) 743 033

Notes

1. Such interest groups included international investors, the state elite, indigenous African and immigrant business society, and white and black large-scale commercial farmers.
2. I use the term 'mainstream' to distinguish these theories from gender-sensitive studies of land, as distinct from using it to denote the economic theory of African land tenure, which is also termed as the mainstream.
3. For more detailed discussion on these theoretical positions, see Kjell Havnevik (1997) and Kaori Izumi (1998i and ii).
4. This argument emphasises the flexibility and adaptability of indigenous land tenure to changing conditions, which is assumed to lead to spontaneous development and to more efficient and secure land tenure.
5. Tanzania initiated its own structural adjustment in 1982 after a failure of negotiations with the IMF/World Bank. A final agreement with the IMF/World Bank on structural adjustment was reached in 1986.
6. Given at the Ministry of Lands, Housing and Urban Development workshop on a Draft Bill.
7. Issa Shivji, who was the chairman of the Land Commission, has initiated a lobbying against the Bill.
8. There are a number of measures introduced under the structural adjustment programme, which contributed to increased interests among urban people and civil servants in sugarcane cultivation in Turiani in Morogoro region. These included, for instance, increased producer price of sugarcane, availability of inputs and spare parts, increased donor supports and new opportunities for private investment, and retrenchment of civil servants.
9. The case in Kagera region was presented by Ambreena Manji at a seminar organised by Women's Front in Oslo, Norway, March 1999.
10. Black population here includes those of Indian descent and of mixed race.
11. Land tenure in Zimbabwe is categorised as urban land; commercial and industrial land; resettlement land; communal land; large-scale commercial agricultural land; and small-scale commercial agricultural land (Gaidzanwa 1995).

References

Agarwal, Bina (1994) *A Field of One's Own: Gender and Land Rights in South Asia*, Cambridge University Press.

Alexander, Jocelyn (1991) 'The unsettled Land: The Politics of Land Distribution in Matabeleland, 1980-1990', *Journal of Southern African Studies*, Vol.17, No.4.

Amanor, Kojo Sebastian (1997a) 'Restructuring Land Relations in Ghana: Adjustment, Capitalism and the Peasantry', preliminary report, Institute of African Studies, University of Ghana.

Amanor, Kojo Sebastian (1997b) 'Collaborative Forest Management, Forest Resource Tenure and the Domestic Economy in Ghana', IRD Currents, Department of Rural Development Studies, Swedish University of Agricultural Sciences, Uppsala, Sweden.

Berry, Sara (1993) *No Condition is Permanent: The Social Dynamics of Agrarian Change in Sub-Saharan Africa*, Maddison, University of Wisconsin Press.

Berry, Sara (1994) 'Resource Access and Management as Historical Processes', in Christian Lund and Henrik Secher Marcussen (ed.) *Access, Control and Management of Natural Resource in Sub-Saharan Africa: Methodological Considerations*, Occasional Paper No.13, International Development Studies, Roskilde University, Denmark

Bruce, John W and Migot-Adholla, Shem E (eds.) (1994) *Searching for Land Tenure: Security in Africa*, The World Bank, Kendall/Hunt Publishing Company.

Falloux, F (1997) 'Land Management, Titling and Tenancy, in TJ Davis and I Shirmer, 'The Seventh Agricultural Sector Symposium', The World Bank, Washington DC.

Gaidzanwa, Rudo (1995) 'Land and the Economic Empowerment of Women: A Gendered Analysis', *SAFERE* Vol. 1 No.1 January 1995.

Government of Zimbabwe (GOZ) (1982) 'The Legal Age of Majority Act', printed by the Government of Zimbabwe, Harare, in Anne Hellum (ed.) (1999) *Compendium of Legal Texts, Cases and Reports, Women's Law and Human Rights*, Institute of Women's Law.

GOZ (1985) 'Constitution of Zimbabwe, As amended at the 1st August, 1985', printed by the Government Printer, Harare, in Anne Hellum (ed.) (1999) *Compendium of Legal Texts, Cases and Reports, Women's Law and Human Rights*, Institute of Women's Law.

Havnevik, Kjell (1997) 'The Land Question in Sub-Saharan Africa', IRD Currents, Department of Rural Development Studies, Swedish University of Agricultural Sciences, Uppsala.

Izumi, Kaori (1998i) 'Economic liberalisation and the land question in Tanzania', Ph.D. dissertation, International Development Studies, Roskilde University.

Izumi, Kaori (1998ii) 'Process and Structure on the Land Question in Africa: Some Theoretical Issues', working paper, Centre for Development and Environment, University of Oslo.

Jacobs, Susie (1996) 'Structure and Processes: Land, Families, and Gender Relations', in *Gender and Development* Vol.14, No.2.

Jacobs, Susie (1998): 'A Share of the Earth? Feminisms and Land Reforms in Zimbabwe and South Africa', proceedings of the international conference on land tenure in the developing world with a focus on Southern Africa, University of Cape Town, 27-29 January 1998.

Johnson, Ootunde E G (1972) 'Economic Analysis, the Legal Framework and Land Tenure Systems', in *Journal of Law and Economics*, Vol. 15, No. 1.

Kanyinga, Karuti (1997) 'The land question and politics of tenure reform in Kenya, the land question in sub-Saharan Africa', IRD Currents, Swedish University of Agricultural Sciences.

Kanyinga, Karuti (1998) 'The land question in Kanya: Struggles, accumulation and changing politics', Ph.D. dissertation, International Development Studies, Roskilde University.

Lastarria-Cornhiel, Susana (1997) 'Impact of privatisation on gender and property rights in Africa', in *World Development*, Vol. 25. No. 8.

Mac Auslan, Patrick (1996) from a presentation at a workshop on a draft Bill for a Land Act held in Arusha organised by the Ministry of Lands, Housing and Urban Development, November 1998, Arusha.

Manji, Ambreena (1998) 'Gender and the

politics of the land tenure reform process in Tanzania', *Journal of Modern African Studies*.

Manji, Ambreena (1999) 'The AIDS epidemic and women's land rights in Tanzania' in *Recht in Africa*.

Meinzen-Dick, Ruth S (1997) 'Gender, property rights, and natural resources', in *World Development*, Vol.25, No. 8.

Moore, Sally (1998) 'Changing African land tenure: Reflections on the incapacities of the state', in Christian Lund (ed.), 'Development and Rights: Negotiating Justice in Changing Societies', *The Journal of European Development Research*, Vol. 10, No. 2.

Moyo, Sam (1995i) *The Land Question in Zimbabwe*, SAPES Books, Harare.

Moyo, Sam (1995ii) 'A gendered perspective of the land question', *SAFERE*, Vol. 1. No. 1.

Moyo, Sam (1997/8) 'The Land Acquisition process in Zimbabwe (1997/8)', *SAFERE*, SAPES Trust, Harare.

Moyo, Sam (1998) 'The Impact of Structural Adjustment on Land Uses in Zimbabwe', paper presented at the synthesis conference on Structural Adjustment and Social and Political Contexts in Sub-Saharan Africa, Copenhagen, 3-5 December 1998.

O'Flaherty (1998) 'Communal tenure in Zimbabwe: Divergent models of collective land holding in the communal areas', *Africa* Vol. 68, No. 4.

Platteau, Jean-Philippe (1996) 'The evolutionary theory of land rights as applied to sub-Saharan Africa: A critical assessment', in *Development and Change*, Vol.27, No.1.

Rwangira, Magdalene K (1997) 'A critical review of the draft Bill of the land act from women's perspectives', a paper presented at the consultative women's workshop on the draft Bill for the Basic Land Act held at the Russian Centre, Dar es Salaam, 3-5 March 1997.

Supreme Court of Zimbabwe (1999) 'Judgement No. S.C. 210/98, Civil Appeal No. 635/92', by Cubbay CJ, McNallyja, Ebrahimja Muchechetereja & Saduraja, Harare, November 2, 1998 and February 16, 1999, in Anne Hellum (ed.) *Compendium of Legal Texts, Cases and Reports, Women's Law and Human Rights*, Institute of Women's Law.

Tanzania Women Lawyers Association (TAWLA) (1997), report of proceedings and recommendations, the consultative women's workshop on the draft Bill for the Basic Land Act held at the Russian Centre, Dar es Salaam, 3-5 March 1997.

United Nations High Commissioner for Human Rights (1995) 'Initial Report: Zimbabwe. 25/09/95. E/1990/5/Add.28 (State Party Report), Implementation of the international Convention on Economic, Social and Cultural Rights', in Anne Hellum (ed.) *Compendium of Legal Texts, Cases and Reports, Women's Law and Human Rights*, Institute of Women's Law.

United Republic of Tanzania (URT) (1994) 'Report of the Presidential Commission of Inquiry into Land Matters, Vol. I: Land Policy and Land Tenure Structure', The Ministry of Lands, Housing and Urban Development, in co-operation with the Scandinavian Institute of African Studies.

URT (1995) 'A National Land Policy', The Ministry of Lands, Housing and Urban Development.

URT (1996) 'A Draft of a Bill for the Land Act'.

World Bank (1989) *Sub-Saharan Africa: From Crisis to Sustainable Growth*, The World Bank.

World Bank (1992) 'Empowering villages to manage their natural resources: Rural land policy in Tanzania', World Bank.

Does land ownership make a difference? Women's roles in agriculture in Kerala, India

Shoba Arun

Women who own land may still lack control over it. Despite claims that women enjoy high status in Kerala, economic, social, and cultural factors interact to reinforce gender differences in ownership, control over, and access to critical agricultural resources, including land.

Since the 1980s, Kerala has received attention because of its combination of low economic growth with high social development, compared to the rest of India and to other developing countries. Gender and development researchers are well-acquainted with its impressive statistics on women: for example, Kerala's female-male ratio is 1,036 females for every 1,000 males, compared to a sex ratio for India as a whole of only 927 females for every 1,000 males (GOK 1997). Life expectancy at birth is 69 years for men and 72 years for women, compared to 60.6 for men and 61.7 for women in India as a whole. According to the 1991 census, Kerala's literacy rate was 89.81 per cent, whereas India's average rate is 52.21 per cent; female literacy was 86.17 per cent in Kerala, compared to the national average of 39.29 per cent (GOK 1997).

Kerala's achievements relating to quality of life, high life expectancy, high literacy, and low infant mortality are due to various social, historical, and political reasons, but three key factors can be identified (Sen 1993). The first is the relative autonomy of the government in two of Kerala's three sub-regions during the colonial period, which enabled it to spend more on health and education, creating public awareness as well as infrastructure. Second, women have been able to get equal access to these services, due to the matrilineal system of descent in Kerala, which has had a great influence on social and cultural development in Kerala. It has contributed to changing social attitudes and created conditions in which women made real progress in health and education.[1] Third, a surge of social and religious reform movements in the 19th and 20th centuries allowed social benefits to spread down the caste hierarchy, and a high level of democratisation. Since 1956, successive governments in Kerala have been instrumental in improving health and education, and have also introduced radical land reforms, relatively high minimum wages, and a wide network of social security schemes (Panikar and Soman 1984). Improving 'woman's agency', for example by promoting female literacy, is seen as contributing much to Kerala's exemplary social development (Panikar and Soman 1984).

However, Kerala currently faces a number of crises, including low economic growth, high unemployment, and a mounting fiscal crisis. Economic liberalisation policies introduced in 1991 have led to fluctuating prices of cash and food crops, and the agricultural sector is in decline. This article examines the impact of the economic changes brought about by liberalisation on Kerala's farmers, who are mostly smallholders, and links this to the questions of women's participation in agricultural production and gender equity in Kerala. The invisibility of women in Kerala's public domain has drawn considerable attention during recent years.

Indian women's need for land rights[2] has been argued for on grounds of family welfare, efficient national development, gender equity, and women's empowerment, and the argument that female inheritance of land would lead to land fragmentation has been refuted (Agarwal 1994). From a welfare perspective, women's land rights are promoted in the belief that they will enable women to have direct access to productive resources, so enabling them to meet their households' basic needs. But I have found that lack of direct access to productive resources is common even in households where women own land — a significant factor in perpetuating not only household poverty and economic inequality between women and men, but also social and cultural inequalities, both inside and outside the household.

The research

The article draws on a 12-month period of research into women's role in agriculture and how this is determined by their socio-economic and cultural context, looking at both matrilineal and patrilineal[3] households.

The matrilineal households (35 per cent of the total households included in the research) were *nairs* (upper Hindu castes). *Nair* women may inherit and own land and property. The *nair* system of marriage, residence, land holding, and inheritance has had tremendous implications for the status of women. Yet *nair* women have had little control over managing property, because men are the official heads of households (Gough 1972, Agarwal 1994). In addition, in 1960, the Kerala Agrarian Relations Act conferred ownership rights to tenants of land, and limited the extent of surplus land held by large landowners, which led to much land being taken away from matrilineal households (Saradamoni 1983).

The patrilineal households were Syrian Christian (30 per cent of households in the research) *ezhava* (lower Hindu caste) and Muslim (10 per cent each), and scheduled castes[4] and other Christian communities (15 per cent). The scheduled castes mostly owned only dwelling plots, although some households leased small plots of land, usually less than half an acre. In patrilineal households, women do not own or inherit land; instead, they are provided with a dowry upon marriage, which may take the form of land given to the bridegroom, gold, or cash, according to the preference of the groom's family. The daughter's family may pay for the non-land portion of the dowry by selling land or property; this is clearly negotiated, usually before the marriage. In most cases, women do not actually have a claim to household property, such as land. This has implications for their access to critical farm inputs and services, and agricultural practices.

Ummanoor and Moorkanad

The research took place in two locations, Ummanoor and Moorkanad. The first has a large *nair* community, while the latter has a large proportion of Muslim households. Communities of Christians, *ezhavas* and *pulayas* are found in both regions.

The *panchayat*[5] of Ummanoor is in the midlands, where the soil and water resources enable farmers to grow a variety of crops, including rubber and rice (GOK 1996a)[6]. Agriculture is the main source of income in this *panchayat*; however, I found many men from farm households employed in teaching and clerical jobs. In spite of the high literacy rates of both women and men, it is not usual for women to take up paid employment, and particularly rare after marriage, when the husband's family may take such decisions. In a small proportion (7 per cent) of cases, I found women had taken up teaching, if the school was located nearby; most of them were getting support from their natal families to meet their household responsibilities. More than three-quarters of land holdings in Ummanoor are below half an acre[7] in size, while 19 per cent are between 0.5-2 acres of land (GOK 1996a). Only 4.4 per cent of the holdings are larger than this. In addition to agriculture and formal employment, cashew nut processing is a major source of income (other activities of lesser importance include brick-making[8]). Owing to the general stagnation of the cashew industry, most factories have moved to other states, thus leading to large-scale displacement of labour.

The *panchayat* of Moorkanad is in northern Kerala, in Malappuram district. The topography of the region is very varied, and crops include cashew, rubber, banana, tapioca, vegetables, coconut, and arecanut. Here, it can be seen that the much-acclaimed Kerala model of development has not fully distributed benefits to the lesser-developed regions in the state. Malappuram district has the lowest income per capita in Kerala: 24 per cent of households are without houses or farmland, though some may own their dwelling plots (GOK 1996b). Out of a total of 3,081 landholders, 82 per cent own less than half an acre, 13 per cent have between half and two acres, and five per cent own more than 2 acres of land (GOK 1996b). Moorkanad is an industrially under-developed area, and there are also inadequate education facilities for higher education. The major medical centre in the area provides primary health care[9]; it is in poor physical condition, lacking a proper water supply, an adequate supply of medicines, and transport facilities (GOK 1996b).

The region has high levels of unemployment, and there has been a significant level of male out-migration[10] for many years, as a large population of this district works in the Gulf. In 1992-3, an estimated 119, 200 migrants left Kerala for various destinations around the world: 53 per cent of these migrated to the Gulf (GOK 1997).

Many women and men in both locations reported that farming is seen as less profitable than it used to be. Over the past two decades, the cultivation of labour-intensive crops like rice has declined in Kerala, and tree crops like coconut or rubber are cultivated instead, since they command a higher market price and require less labour. However, my respondents reported a steady increase in prices of inputs like fertilisers, and declining and unstable crop prices. Within six months in 1996/7, the price of one coconut varied between ten and two rupees, and the prices of rubber fell from 48 rupees per kg to 25 rupees per kg, resulting in extreme income instability (GOK 1997). For many households, income from agriculture is now seen as a source of additional income rather than a livelihood in itself. An increasing preference for men to find employment outside the household has important implications for farming and gender relations in both matrilineal and patrilineal households.

Social norms and the gender division of labour

Overall, 45 per cent of households in the research had a man in paid employment in the formal sector, while 30 per cent had a man engaging in informal sector work, for example as an artisan or driver, and 15 per cent had a man in paid farm labour. Almost 48 per cent of women were managing the family farm, because their husbands had paid employment, were migrants, or absent for another reason. About 35 per cent of women were involved in paid work, and 7 per cent of women were engaged in paid employment in the formal sector like teaching, but also undertook some of the farm supervision. In households where men migrate to the Gulf (47 per cent in Moorkanad), women shoulder a particularly heavy workload. It is amazing to see women single-handedly taking on all the responsibilities of the financial and social organisation of the farm household, combining productive and reproductive tasks.

In both locations, and in matrilineal as well as patrilineal households, women's limited role in many agricultural activities and traditional, narrow understandings of women's work led to constraints on what women were able to do on the farms[11], and to a lack of recognition of women's contribution to the household: for example, many women who manage the farm considered themselves to be housewives. This lack of recognition affects women's control over how income is spent, and their authority to participate in decisions regarding the sale of land or transfer of control or ownership of land to other family members.

Changing relationships: Women and land

Nair women's control of land

Bina Agarwal argues that, in South Asia, women's land rights can, over time, help women negotiate less restrictive norms and better treatment from husbands (Agarwal 1997). Although, in *nair* households, matrilineal inheritance laws enable women to inherit property such as land or houses, ownership of land does not seem to translate into control over it, or the income from it, and improved power relationships in the household. In many of my sample of *nair* households, women stand to inherit a majority of the family property, including farmland and other resources, but their ability to continue farming is largely determined by factors such as post-marital residence and paid employment. In several *nair* households, I found that the husband had taken charge of the household and farm responsibilities after marrying and moving to his wife's home.

Nair households used to be matrilocal[12] as well as matrilineal, but I found that this has changed for many households. Women who live with their husbands away from their natal homes, may see their share of property sold if it proves inconvenient or unprofitable. In the current economic climate, selling land is an attractive option for many, and the income accrued is used to buy property or build elsewhere. This process is given legitimacy by ideas about male control of dowry: a woman may inherit some of the assets of her natal family like property or gold, but the husband may use it as capital for setting up business, buying more land, or building a house. Revathi, aged 36, who described herself as a housewife (in the sense that she had no formal paid employment) told me:

'Sthridhanam (dowry) is given to our husbands, not to us, so that he can look after us. He and his family have all the right over it, whether it is cash, gold or land.'

Another woman considered farming as not so lucrative as other business. She told me:

'Women are brought up to support decisions which are best for the family ... for instance, I feel sad to sell my family property, but for the sake of a better future for my children, I have to do it.' (personal conversations, 1996/7)

In other cases, *nair* women have moved away from their land to accommodate a husband's paid employment — either to his home or a third location — and the wife's natal family is controlling her share of property, although the husband also has customary rights in matters over his wife's property. In many cases, the husband or the natal family seem to have more control over the woman's share of property than the woman herself. These findings in Kerala echo a study from north-west India, where many women who inherited land had only minimal control over the land they officially owned (Sharma 1980).

Nevertheless, in the Kerala study, some *nair* women were able to retain control over land and property which they inherit. For example, in some households, where husbands fail to do their share in maintaining the household, or do not save for the future, women — often with the help of their natal families or close friends — save the income accrued from cultivation on their share of the property, including that from cash crops such as rubber.

Changes to agricultural practices

Social norms regarding women's work, and women's need to combine caring work with agriculture have led to changes in agricultural practices. The fact that labour-intensive crops such as rice and tapioca are being replaced by tree crops which need less attention is positive. However, as one woman stated, 'often social norms restrict my time and labour as I cannot engage in cultivation like a man'. Many women in my study were supervising the farm, but avoiding tasks that are considered 'men's work', such as harvesting rubber and coconut, and purchasing agricultural inputs. Men were hired to take on these tasks.

When men are engaged in paid employment locally, both men and women can make decisions about the farm, although men are the primary decision-makers in most households. In contrast, in households where husbands are absent, women are left to make decisions themselves. Some told me they were unhappy with this big responsibility, especially with tasks like organising labour. Some women had decided to reduce the extent of farming and concentrate on growing most crops in the homestead garden. Another solution is to seek the assistance of close male relatives for labour, but depending on them leads to their involvement in household responsibilities and decision-making.

Wives do not retain income from the main crops — this usually goes to their husbands. However, women do usually keep the income from secondary crops, such as cashew and tamarind, and from garden crops. In some households, I found that income derived from garden crops was being used by women to buy household appliances, to save in *chitty*[13] (informal savings) as a contingency fund, or for purchasing gold for their daughters.

Accessing extension services

In this context, women's ability to function as independent farmers needs to be enhanced by policies which support them in gaining direct access to credit, production inputs, information about agricultural practices, and which rectify the male-biased farming system.

In both locations, the Kerala state government has developed agricultural programmes, administered by local farm offices in order to distribute improved seeds and plants, pest management systems, and assist in small-farm mechanisation. But both male and female respondents in my research reported that they do not receive adequate and timely information that is critical for farm productivity.

In my conversations with women, it was clear that women have a great deal of indigenous knowledge about farm activities and crop cultivation, and, in the absence of exposure to technical knowledge which could potentially be gained from extension services, they resort to traditional methods of farming. However, women feel that they are less able than men to access technical advice. For example, Ummanoor women told me that farm extension services in their area tend not to approach women, because it is assumed that 'farmers' are men. During my field work in early 1996, the incidence of root wilt for coconut was widespread, because farm extension officers failed to approach the households' women to discuss the scientific management of the problem. Similarly, most of the Moorkanad women who are left alone to head households while their husbands are away told me that they never approached the local farm offices (*Krishi bhavan*) directly, owing to a combination of social inhibitions and their increased involvement in other household activities.

From my own observation, the extension officers — who are after all assigned to disseminate knowledge — only visited the fields, rather than searching out women in their homes. Although the social impropriety of visiting women at home might be cited as a reason, it is often assumed that women have no significant role in farm activities. Many women listen to agricultural programmes on the radio and television, and read daily newspapers, in an effort to improve their agricultural practices. This not only proves women's awareness of the need to improve agricultural methods and inputs, but is also a telling example of the instrumental value of women's literacy as a potential tool to increase farm efficiency and to ensure women's participation in the development process.

An example of how women who farm are marginalised from essential information is the wilting disease which, in 1995, afflicted a new high-yield pepper, Panniyur-I. In Moorkanad, the disease destroyed almost all pepper cultivation. Women, who had little land, labour, or money to hire labour, suffered most, since they had concentrated on growing garden crops which do not need much labour and attention. Most women I spoke to had attributed the disease to natural causes, and did not receive timely information about the disease or alternative varieties; to a large extent, they had to rely on secondary information on agriculture gained through family networks, and other local farmers. They had also not known at the time that they could have claimed for compensation (personal communication, 1996/7). This type of exclusion affects farm productivity and income: the prices of pepper had increased over 99 per cent between 1991-97, especially after the introduction of economic reforms (GOK 1997).

Accessing banking and markets

In Moorkanad, women also complained of lack of access to banking and credit facilities. In several households, women wanted to expand the farm or install pumps for irrigation, but said that they had been refused a loan because they could not provide any collateral security. There are only two banks in this region, and informal lending for high interest, or pawning of gold and other assets, are the main ways in which women borrow money; because they lack collateral in the form of land or production equipment, most are unable to borrow from formal institutions like banks, which could give them better rates of interest. Lack of access to credit also means that women have little flexibility in choosing income-earning activities to embark on without the help of male household members, who own the assets that could be used for collateral security.

Moorkanad's marketing facilities are also inadequate (GOK 1996b): for instance, despite increased production of coconut over the past 20 years, the oil mills in the region can only extract coconut oil for household purposes, lacking the capacity to meet the demands of commercial production. Farmers who cannot afford to transport unprocessed coconuts to markets out of the area, or who are unable to travel in order to do so, must sell them to local agents for a lower price, which creates a crisis for producers due to the decline in prices mentioned above. In this, as in other situations, women have to face the reality of the market's discrimination against them (Harriss-White 1995). In the words of one woman farmer in my study: 'I cannot bargain with men in the market; I have to oblige with the social norms in our society'.

Decreasing opportunities in agricultural labour

In my sample, there were eight households across the two locations who were Latin Christians and belonged to the scheduled castes. Being engaged mostly in agricultural work, they were largely landless, although some owned their house plot. As noted earlier, women from Christian families do not inherit land, but bring a dowry in the form of land endowed on their husbands, cash, or gold. In this category of household, women are disadvantaged at several levels.

Agricultural wages have increased nearly nine times for men, and over eight times for women respectively over the past 16 years in Kerala (GOK 1997), yet opportunities for paid work have decreased due to the transition from labour-intensive rice cultivation to less labour-intensive crops including rubber and coconut. Agricultural workers, especially from the lower castes like the *pulayas* and 'backward' Christian communities, usually cultivates crops like rice, tapioca, and vegetables which fetch lower prices compared to rubber and coconut. Since they work on very small land-holdings (whether of their own or leased), typically of about 0.2 acres in size, it is not practical for them to cultivate rubber which takes a long time (about seven years) to grow. They thus continue to cultivate other, minor, crops, although they do not make a profit. Many male former agricultural workers have moved to take up artisanal jobs like carpentry, which provide more income and have a higher status than farm labour. One woman agricultural worker told me: 'I have to plead for higher wages to earn more so that I can feed my children'; this woman was also cultivating a small family farm.

Conclusion

As illustrated in the above examples, women may gain access to land in numerous ways — through inheritance, marriage, or informal networks. However, none of these options guarantees effective command over it. Women's traditional rights to land have not been adequately recognised in Kerala: the gender gap in the ownership and control of property is the single most important contributor to the gender gap in women's economic well-being, social status, and empowerment (Agarwal 1994). It is disquieting to note that the current socio-economic changes and crisis of confidence in agriculture as a main source of livelihood is leading to nair women's share of land being sold, with the proceeds going to men — thus reducing women's ownership of land to the status of male-controlled dowry.

Although legal provisions such as equal access to employment or land are important in recognising women's rights to land, legal rulings alone can have limited impact on changing gendered power structures within societies, families, or communities. The case of the Christian succession laws in Kerala is a case in point. In 1986, a group of pioneering women questioned the validity of the Travancore Succession Act of 1926, which gave property rights to sons rather than daughters. Although a favourable verdict was obtained from the Indian Supreme Court, this was strongly opposed by the church, state, and other institutions, and led to the ostracism of these women (George 1994).

Land is the most basic resource of agricultural production. The recognition of women's differential access to property and their lack of command over its use — even if they own it — should be the starting point of a gender-sensitive agricultural policy. Since 1996, Kerala's government services have been undergoing a process of decentralisation, with the aim of enabling more participation at the local level. However, women in agriculture have yet to be incorporated effectively at local level. Women involved in farming are hampered from gaining access to credit, extension, and marketing, in addition to the purchase or lease of land. Women's role in farming in Kerala needs to be recognised, and institutional support must be increased, to enable women to gain access to agricultural inputs and technology, which would lead to better agricultural practices and a higher income from farming.

Increasing male migration for work and diversification into paid employment means a growing number of de facto female-headed households, where women are wrongly perceived as dependent on men, despite their primary — and growing — responsibility for the daily financial and management and organisation of the household and the farm. It is important that women have direct access to critical farm inputs, to enable them to maximise outputs, challenge ideas of 'women's work', and thence to gain control over the other factors of production and change social norms. Most importantly, there should be a concerted effort to enable women to function as independent farmers who control their own land.

Shoba Arun has recently completed her PhD, on gender and agricultural households in Kerala, India, at the University of Manchester. Contact details: Department of Sociology, Williamson Building, University of Manchester, Manchester, M 13 9LW, UK. Tel: +44 (0)161 273 4612; e-mail: s.v.arun@stud.man.ac.uk

Notes

1 The matrilineal system of inheritance and family organisation (practised by the *nairs*, the *ambalavasis* and by sections of the *ezhava* caste) involves property being inherited by and through women; a typical household consists of a male head, his sisters, and their children and grandchildren, while husbands have 'visiting rights'. Matrilineal, however, does not mean matriarchal: women do not dominate in household decision-making power. As this article will show, matriliny may have given women importance, but this has stopped far short of equality with men (Jeffrey 1993).

2 Agarwal defines land rights as 'rights that are formally untied to male ownership or control, in other words, excluding joint titles with men. By effective rights in land I mean not just rights in law, but also their effective realisation in practice' (Agarwal 1994, 3).

3 Patrilineal household: where inheritance is passed down the male line.

4 *Ezhavas* are the lower caste Hindus who along with Christians and Muslim population used to be largely involved in trading occupations, while the *pulayas* and the *cherumas* are the scheduled castes who are largely the deprived classes as they were regarded as the 'untouchables' in the society until the early twentieth century.

5 Kerala is divided into 14 districts and 1,000 *panchayats*, which is the smallest unit of local administration.

6 According to the Agro-climatic Committee of 1974, Kerala has been conceptually divided into five agricultural zones based on the characteristics of rainfall, climate, soil, topography and elevation (NARP 1982).

7 One acre is equal to 0.25 hectares.

8 Brick-making is a highly seasonal job, and dependent on weather conditions.

9 A primary health-care centre serves a about 30,000 people and caters for the health needs of the local population, especially in family planning and immunisation services. It is further supported by a community health centre which covers a population of 100,000, and a sub-centre covering four to five villages.

10 Kerala contributes to nearly 50-60 per

cent of Indian workers in the Middle East (GOK 1997). This outflow has increased since the beginning of the 1970s on account on the hike in oil prices and large-scale investment in all the oil exporting Arab countries. Although a large number of skilled and educated persons regularly migrate to countries like the USA and the UK for work and education purposes, the pattern of migration to the Gulf is quite different. First, it is mostly temporary, as well as circulatory and repetitive. In addition, most of the migrants are unskilled and uneducated, and work as manual workers. There is a large predominance of males among these out-migrants, as most families cannot take their families on account of inadequate income and also owing to the constraints placed on women's migration by the Emigration Acts of India.

11 Each household organises and manages agricultural activities and household responsibilities differently, depending on various factors, including norms about the division of labour between women and men. These norms may restrict women who manage the family farm from engaging in certain agricultural tasks — for example, women do not operate machinery such as tractors, tap the rubber trees, or climb trees to harvest coconut and arecanut.

12 Matrilocal implies that on marriage, the husband moves to his wife's family home.

13 *Chitty* is a form of popular informal savings among households in southern India. On the initiative of one person, a group of individuals — say, five persons — pool equal sums of money — say, Rs 200 per month. On one day of every month, there will be a raffle where one person's name is drawn who receives the whole amount of money i.e. Rs 1,000. In this way, every month one person is entitled to the total money.

References

Agarwal, B (1994) *In One's Own Field*, Cambridge University Press.

Arun, S (1999) 'Gender, agriculture and development: The case of Kerala, South-West India', unpublished Ph.D. Dissertation, University of Manchester.

George, Ajitha (1994) 'Kerala Syrian Christian women and subordination', AKG Centre, Trivandrum.

Gough, K (1973) 'Kinship and marriage in south-west India', in *Contributions to Indian Sociology*, No. 7.

GOK (1996a) 'Adhikaram janagalilekku', Ummanoor Panchayat report, Kerala.

GOK (1996b) 'Adhikaram janagalilekku', Moorkanad Panchayat report, Kerala.

GOK (1997) 'Economic Review', State Planning Board, Trivandrum.

Jeffrey, R (1993) *Politics, Women and Wellbeing: How Kerala Became a Model*, Oxford University Press, New Delhi.

NARP (1982) 'National Agricultural Research Project', Kerala Agricultural University.

Paniker, PGK and Soman CR (1984) 'Health Status of Kerala', mimeo, Kerala, Center for Development Studies.

Saradamoni, K (1983) 'Changing land relations and women: A case study of Palghat District, Kerala' in V Mazumdar (ed.) *Women and Rural Transformations*, Concept Publications, New Delhi.

Sen, Gita (1992) 'Social needs and public accountability-The case of Kerala' in Wuyts, Marc, Maureen Mackintosh and Tom Hewitt (eds.) *Development Policy and Public Action*, Open University Press, London.

Sharma, U (1980) *Women, Work and Property in north-west India*, Tavistock Publications, London.

Rural development in Brazil: Are we practising feminism or gender?

Cecilia Sardenberg, Ana Alice Costa, and Elizete Passos

In the past two decades, the concept of gender has become central to feminist scholarship and activism. It is a powerful instrument towards the empowerment of women, but with the mass use of the term, its political meaning is being lost, which may lead to women becoming invisible once more.

At least in theory, gender awareness means greater visibility for the way in which development planning is shaped by patriarchal social relations. Planners are obliged to consider the social, political, economic, and cultural forces that determine women's and men's control of resources and products and therefore the degree to which they can participate in — and benefit from — development efforts. However, feminists working in development are currently concerned that the widespread use of the term 'gender' by mainstream development planners is 'contributing to its vulgarisation and simplification' (Celiberti 1997, 69), and the effacement of its political meaning (Costa and Sardenberg 1994).

In this article, we consider these issues, while sharing our experience as gender consultants on a state-sponsored rural development project in an arid area of Bahia, Brazil, where agriculture is one activity in the complex livelihood strategies of poor women and men. We shall refer to the project as the Eagle River project, and to the target area as the Eagle River region. The project is a major development effort in one of the most deprived and poorest regions in north-eastern Brazil, and is the first one there which incorporates a gender analysis.

Implementation of the project began in December 1997, but only six months later the implementing agency, the Regional Agricultural Development Company (CAR), an agency linked to the planning bureau of Bahia, contacted us to devise a gender programme.[1] We work as external consultants through REDOR, a regional network of women's studies centres which includes our own institution, Núcleo de Estudos Interdisciplinares sobre a Mulher (NEIM), the Centre for Interdisciplinary Women's Studies of the Federal University of Bahia. NEIM was founded in 1983, as the second women's studies centre in Brazil. We should state at the outset that we are self-proclaimed feminists, active in women's movements in Brazil.

The Eagle River project

The Eagle River project is an ambitious undertaking, covering a large area of one of the least hospitable regions of Bahia. Its implementation was planned to take five years, with the aim of developing an area encompassing 13 counties covering about 4,580 square miles, in the centre-south region of the state. This area consists almost entirely of *caatinga* vegetation,[2] with several distinct ecological sub-regions, some of which are considerably more humid and more fertile than others. However, agricultural production is limited due to lack of water: the area frequently suffers from long seasonal droughts. A drought can last as long as seven years, since the rain is never adequate to replenish water holes or rivers. Part of the project area is located in an area called the 'drought polygon', and one county has the lowest annual rainfall in the whole of Bahia. Although there are waterways throughout the area (including the Eagle River itself, which flows through all the 13 counties), most rivers are seasonal and almost completely dry up during the drought periods. Other water resources such as dams and waterholes are scarce and unevenly distributed, and access to them is difficult for many local producers.

As a result, the region has not attracted major agricultural and industrial enterprises. In contrast to most other regions in Bahia where large estates predominate, the Eagle River region is characterised by small land-holdings (*minifundia*). Close to two-thirds of the population live off the land, staying close to their relatives in small communities. Most *minifundia* are less than 100 hectares, and many are under 10 hectares. Nearly 85 per cent of the properties are worked by the owners themselves with the help of family members; the other 15 per cent regularly hire outside hands, or work as hired hands on other people's properties. The larger properties are usually owned by people who live in the city and who employ hired hands to run their farm. Those who run theses places usually do not own land, or own such small areas that it is not worth the trouble to farm them.

Due to the adverse ecological conditions, the small size of the properties, and the lack of available credit to small farmers and peasants in the region, production returns tend to be low, barely covering family needs. The vast majority of the properties are geared primarily to subsistence production and the sale of surplus products. Families in the region plant corn, beans, manioc, and sugarcane, and also raise cattle and small farm animals. Almost 65 per cent of rural families in Bahia live in poverty (averaging an annual income of less than US$2,500 per household (internal surveys from Secretaria de Planejamento e Tecnologia (Seplantec)/CAR, undated). Less than 35 per cent of the rural communities in the project area have electricity, and only a few have a local health facility with an attending nurse. A family has between five and six children, and child mortality rates are high: 88.20 per 1,000 (ibid.). Rural elementary schools (taking children up to the fifth grade[3]) have only been set up in recent years, together with transport to enable older children to attend high schools in the main county-towns. At 49 per cent of the population[4], the area's illiteracy rate is one of the highest in Bahia.

The county towns are the only urban areas; for the most part, they are small in size and unimpressive in terms of services and commercial activities. One of the few major events in these towns are the *feiras* (weekly markets or fairs), where local farmers — including women — bring garden produce and small animals for sale.

Women's lives in the project area

In the Eagle River region, as in other rural areas across Brazil, traditional values regarding the division of labour, women's domestic roles, and gender hierarchies still predominate. Women marry young and are entrusted primarily with the care of children and other domestic activities. However, they are also expected to 'help' their husbands in the field, care for small animals raised on the farm, and prepare manioc flour and cheese for home consumption and for sale in the fairs; some also raise produce for sale. Since much of women's agricultural work is performed as part of their duties as wives and mothers, their productive roles and crucial contribution to household survival are largely 'invisible' and unvalued. Even the women themselves tend to undervalue their participation in production: only 25 per cent of the female workforce are officially recognised as rural workers (Instituto Brasileiro de Geografia e Estatístic/ Pesquisa Nacional por Amostra de Domicíli 1996) and covered by social security benefits.

In comparison to women in urban areas, rural women not only have less access to education, health-care, and employment opportunities, but are also much more dependent upon the men in their families and caught in a structure of more unequal gender relations. Indeed, in line with other poverty-stricken rural areas throughout the world, patriarchal family and social structures deny women real rights to land, limit women's access to and control over the proceeds of their own labour, and constrain their decision-making.

Agriculture is not the only source of income: migration to the south, especially to São Paulo, is a major strategy for most local families to supplement their incomes. Although recent statistics on migration rates in the region are not available, the fact that nearly every family contacted thus far by the project's field staff has one or more members working in São Paulo or elsewhere in the south shows the scale of the livelihoods problem in the area. Young women may go alone, or young couples migrate together; some married women with children also head south to work as domestic servants, leaving the care of the home and children to their husbands and fathers. However, this is more likely to happen if the men are unable to find jobs, or are either too old or too sick to migrate, and if there are no young men in the family to take their place as providers. Usually, husbands and adult children migrate, while women and youngsters stay behind caring for the farm. In April, around the start of the dry season, busloads of migrant men leave the area to find jobs as construction workers in São Paulo. Many return in November, when the rainy season is supposed to start, to clear and till the land for planting.

Depending on precipitation rates and yearly harvests, men's migration between São Paulo and their homes in the Eagle River region may last for many years, giving rise to the phenomenon of 'drought widows', and a large percentage of female-headed households. For some, migration is permanent: the small size of the landed properties and large family sizes means there are too many people who inherit too little land. This, added to the adverse ecological conditions, pushes young men and women to go south permanently.

During our work, we found that the loss of labour through migration can effectively double women's workload. Women from one of the communities told us:

'When the man of the house is away, the woman of the house becomes man and woman of the house. That is why I pray for rain. It is too much work on the woman when we are alone. What happens when my husband is away? Well, let me tell you, it is work, work, work. I much rather go to São Paulo myself to clean other people's latrines than stay behind here. I can work all day as a maid in São Paulo and still not work as much as I work here. Besides, there I get paid, here I don't.'

In certain communities, women hold responsibility for nearly 80 per cent of the households and the care of the land for most of the year (internal project document, 1998).

Aims of the project

The main objective of the Eagle River project is to enable the rural population to stay in the area, through raising production levels and improving the overall social and economic conditions in the region. The project aims to achieve this through the following activities.

- Promoting and strengthening local producers' associations. Some communities already had associations; in others, they are being created with the help of the project. Project plans call for work with 'interest-groups of producers', rather than individual families. All activities are to be carried out with groups, in line with principles of community development.

- Developing and implementing services to improve the productive capacity of small properties, including constructing dams and waterholes.

- Offering technical assistance to producers appropriate to local climatic and agricultural conditions, and providing credit to stimulate productivity.

- Improving social conditions through road-construction and maintenance programmes.

- Devising means of water storage for the use of local families.

- Supporting alternative rural education programmes.

- Offering technical and financial support for the implementation of small-scale irrigation systems running to small properties.

- Promoting the development and marketing of products which can take advantage of existing markets, to bring higher returns to local farmers.

To achieve these aims, the project has three main areas of work: community development (based on a commitment to community participation); production development; and provision of rural credit.

The project's structure and decision-making powers

The project area has been divided into four sub-regions, each with a local office staffed by a male agronomist/co-ordinator, a female social worker, and between three and six agricultural technicians, all but one of whom are men. All the technicians have responsibility for a particular county in the sub-region, and over the past year, each has worked with eight to ten communities there. The number of communities involved in the project should double within the next year, and there are plans to hire more technicians to share the work. Most of the technicians are natives of the Eagle River region, who trained in Agricultural Family Schools[5], which offer the equivalent of a high-school education. Overseeing the work of the local offices is a regional office, with a male regional co-ordinator and three monitors (two agronomists, who are male, and one social worker, who is female). The monitors are currently expected to visit the different sub-units throughout the week, although a decision has just been taken to halt this activity. The regional social worker is primarily responsible for communicating the work of the project's local social workers.

The balance of women and men in the organisation changes at the senior level. The Eagle River project's head office is in Salvador, Bahia's capital city. Its head co-ordinator is a woman, as are two of its four area co-ordinators — a woman sociologist (responsible for community development), a male agronomist (in charge of production development), another male agronomist who oversees the building of dams and waterholes, and a woman economist in charge of evaluations. The staff also include a male accountant and a female secretary.

At first sight, the presence of three women in top positions may suggest that women have considerable power in high-level decision-making in the Eagle River project. In fact, this is far from true: the project is administrated by CAR, which is headed by a man, and he has the final word. The balance of power is even more skewed at lower levels of the project: in particular, the regional social worker is often excluded from decision-making, and even from basic activities. On occasion, meetings have been held without her, and she has been told that 'technical matters', which she would not understand, were discussed. She, and the other social workers in the project, are the only field staff who do not have direct access to transport. Women are thus dependent on the male staff to carry out their work.

The project's first 18 months

During the past 18 months, much of the work has been to identify and survey rural communities in each of the counties, mobilising the producers and working with them to draw up a 'community operative plan'. Perhaps the greatest effort has been concentrated on the construction of 95 water holes throughout the Eagle River area, and the development of field centres (named CATS) which offer communities the chance to participate in the development of alternative technologies, and see them in action. Several 'field days' have been held, in the field centres or at producers' homes, to train them in the use of new methods of appropriate technology, and disseminate information about alternative crops. For example, a type of watermelon is being promoted which can be used as food and water for animals, and is drought-resistant for up to two years. Other crops for cattle-fodder are promoted, which are more appropriate for the ecological conditions of the area, and which could serve as 'strategic reserves' for the drought season.

Some of the CATS have been very successful in terms of agricultural output. In one of the sub-units, enough water-melons were produced to distribute to all producers in the community. They kept some of the seeds for their own use, but there was enough to distribute them in other communities as well. However, there has been much less success in terms of women's participation. In the beginning, very few women participated in the field-days, and most of those who did attend were there to cook for participants.

Challenges of incorporating gender into the project

We were asked to respond to the need to involve women in the project, by devising and implementing a gender programme. There are few people in Bahia trained to do the kind of work we do, and we had many misgivings about taking on this responsibility. As members of NEIM, we had often worked as gender consultants for government agencies and in development projects, and a major drawback was the fact that government-sponsored projects are often used for political ends. The present government of Bahia is conservative, whereas our work focuses on raising consciousness of gender in the context of wider social rights. We made it clear from our first contacts with the Eagle River project that our main goal in developing a gender programme for it was to contribute to women's empowerment.

Our knowledge of rural life was mostly academic: we had been involved with a range of extension programmes over many years, working with women's groups, but these were concentrated in urban areas, primarily in the poor neighbourhoods of Salvador. What did we have to offer a rural development project? We were also told by the head co-ordinator for community development, who guided our work, that much of our work was likely to be gender-awareness training for the project staff, mostly male agricultural engineers and technicians, many of whom were natives of the project region. We had no experience in training men in gender-related matters outside an academic setting, and no idea of how we would go about this with men from rural areas. We were afraid that they would be very conservative: Brazilian men are known for their *machismo*. How would we go about sensitising them to gender issues?

Developing the gender programme

We were first contacted by the community development co-ordinator on the project, to whom we report. She asked us to write a proposal for 'doing gender', but what she had in mind at first was a programme for 'women only'. We insisted from the outset that it is necessary to mainstream a full analysis of gendered power relations in all the planned activities of the project. Thus the first document we submitted as gender consultants was geared towards providing a critical gender analysis of the project. This analysis included a point-by-point discussion of the project's components, and how one should proceed in order to guarantee a gender perspective (internal document, 1998). Underlying this analysis was the notion that guaranteeing gender equity implies pursuing two lines of action simultaneously: one that tends to the practical needs of women (Moser 1989), and one which is geared to their strategic needs (ibid.) — that is to say, women's need to challenge the unequal balance of power between women and men.

The programme we have devised aims to meet both sets of needs, so that women in the Eagle River region can participate on more equal terms with men, and draw greater development benefits than they would otherwise. In particular, the programme focuses on the following:

1. widening and increasing women's participation in activities related to technical assistance and training in agricultural and husbandry technologies, as well as to the appropriate use of soils and water resources;

2. guaranteeing women's access to productive resources such as credit systems, water holes and irrigation systems, and legal ownership of land;

3. guaranteeing greater gender equity in community associations and local decision-making structures.

One key activity is to form 'production groups' geared to income-generation, in which women receive technical training, parallel to participating in monthly held gender awareness workshops. In order to create the necessary conditions for achieving the proposed goals, the gender programme also includes specific activities such as providing gender-awareness training for the project staff and those of the partner institutions and agencies involved, as well as a programme of gender-sensitising workshops for local women in leadership positions, and one for school teachers.

The original plans envisaged one production group per county, in rural communities where the field staff were already involved. Staff were asked to identify first any pre-existing women's groups, independent of their nature (women's associations, income-generating groups, and religious groups), and, second, communities where such groups were either just starting, or where the community showed an interest in and potential for starting one. We visited all communities thus identified, and selected 15: one in which there was a goat-keeping women's collective; two where women had well-developed income-generating activities, but pursued them individually, with no history of formal association; two where there were long-established women's groups, but none geared to income-generating activities; and ten where there were strong women's networks, either kin-related or associated with local Catholic pastoral activities, which showed significant production and organisational potential. In total, these groups include about 450 women; the number of participants varies from 15 to 60. Their age, marital status, and level of formal education also vary widely.

Learning from our experience

What does a 'gender programme' involve?

The greatest difficulty we have encountered has been arriving at a common understanding with the senior management of the Eagle River project of what a gender programme should involve, and whether 'doing gender' was the same as 'doing feminism'. Senior staff in the project initially saw a gender programme as merely creating income-generating opportunities for women, primarily by forming women's production groups. The head co-ordinator in particular saw women's economic participation as the only gender issue to be addressed. She insisted on the creation of these groups when we presented our first proposal, although she later agreed with us that there was a need to involve women in the other project-related production activities planned.

An opposing view was — and still is — held by the agronomist responsible for the production-development work. He attended a gender-training workshop held by the international co-operation organisation that co-sponsors the project, before we became involved with the project. At the workshop it was (correctly, albeit simplistically) stressed that 'sex is not equal to gender' and that 'gender is not equal to women'. He holds the view that no special programme should be carried out for women, and that 'gender' is concerned with men and women. However, he has not taken on the idea that a gender analysis is founded on acknowledging the unequal power relations between women and men, and is therefore oblivious to the need for 'empowering' women. He proposed that women be 'incorporated' in all the existing planned activities and programmes; yet, given the asymmetrical character of the pattern of gender relations in the region, women cannot participate on equal terms.

We have been adamant that gender cannot be a mere 'sub-component' in the project, to be contained in women's production groups; nor can it be assumed that 'including' women in all activities will suffice, unless unequal power relations are challenged. For example, most of the workshops with producer groups have involved games and dramatisations and have had a playful, relaxed tone. Our initial focus was to assist the groups in defining their specific production interest, and what was necessary to develop it. At the same time, we used techniques to promote group solidarity and organisational skills. We began a series of monthly gender-sensitising workshops on specific topics. Topics already covered include gender roles, women's organisations and struggles, women's rights, and women's health; the last two will focus on education and women's work. Based on 'gender pedagogy' methodologies (Büttner et al. 1997) which are themselves adaptations of techniques devised in feminist consciousness-raising groups, these workshops build upon women's individual experiences and practical knowledge, to achieve a collective reflection on gender relations and women's condition, and ways of improving the situation.

Although the community workshops have been an enriching experience for all involved, including ourselves, there have been some emotional moments, in particular when the issue of domestic violence is raised. This topic has also been the focus of considerable disagreements between us and some of the local co-ordinators, who insist that the issue of domestic violence falls outside of project goals, and that discussing it with the women may result in the loss of support to the project on the part of the men in the communities. They have taken the matter to the head co-ordinator, who appears to agree with them. We do not consider it to be adequate to raise women's self-esteem and promote their participation in production activities, without 'tampering' with the existing pattern of gender relations in the region, especially as far as domestic violence is concerned. Staff say that when we address this issue, we are 'doing feminism', rather than gender, which should be avoided, even when the women themselves have identified violence as one of the main problems they face.

Support and resistance from men

As stated above, we were pessimistic about men's reaction to our activities. However, we were surprised by the genuine openness and interest shown by the male agricultural extension workers who attended the first staff gender-training workshop in October 1998. Contrary to our initial fears and expectations, they showed a high level of perception of and sensitivity towards the unequal relations between women and men in the areas covered by the project, and made a significant contribution to formulating a concrete action for the gender programme. As natives of the area, they were particularly helpful during our first field trip, which followed the first workshop, pointing out nuances in gender relations in the communities visited which we might have missed otherwise. In a subsequent workshop, they not only took an active part in all the discussions, but also added a special touch to the success of the event. For the 'finale' of this workshop, they presented us with a play in which, dressed as women, they showed in a funny yet poignant matter the economic and social problems faced by women in the region, and how they hoped the gender programme could effect changes in the existing situation.

However, none of the men in the upper echelons of the organisation, or of the partner institutions involved in the technical aspects of production development, participated in the second workshop. Instead, they attended a training workshop on the construction of dams, which, against our efforts, was scheduled on the same date. This reflects the tendency to separate the 'social' from the 'technical' activities of the project. With a few exceptions, these men have shown the greatest resistance to the gender programme, even if at times in a veiled manner.

To assist them to change, we have held monthly workshops in each of the project's local units, when we not only discuss some of the more theoretical and methodological aspects of the gender programme, but also evaluate all the activities underway from a gender perspective, regardless of what component they belong to. This has opened the way for a rich exchange, in terms of looking more closely at the difficulties encountered, as well as at the different forms (and degrees) of resistance. These workshops have given an us opportunity to monitor, month by month, the growth of gender awareness among the staff. We have found that the higher the men's status in the project, the greater their tendency to resist our efforts. Most of the problems we have faced have been related to sexism on the part of the local co-ordinators.

Women's participation in decision-making: How far does this help?

From the above, it can be seen that the fact that a senior manager of a development project is a woman does not guarantee that it will be informed by a feminist commitment to equality between women and men. While it is important in itself to have equal numbers of women and men staff at all levels, and particularly in senior management, only the individual commitment of staff to feminist ideals will actually ensure that the project benefits women through challenging oppressive gender relations.

For many women, 'making it to the top' is accompanied by adopting values and attitudes associated with male managers. In one meeting, the head co-ordinator told us that all that is necessary to improve women's condition is to give them better economic opportunities. This position is at odds with feminist views that women's experience of poverty has social and political dimensions as well as economic ones. She also espouses the idea that 'doing gender' should not be the same as 'doing feminism'. She told us that the project should not 'start a feminist revolution in the area', and that domestic violence is a private issue, which the project should not meddle with.

In contrast, the other women senior managers at head office have made it clear that they sympathise with feminist ideals, agreeing that we must work towards raising women's consciousness of their oppressed situation. They also agree that gender must be mainstreamed in all project activities, but they are responsible for components which are regarded as 'less important' than those co-ordinated by their male counterparts, and are often excluded from the overall decision-making process.

Timing of staff training

As Robert Chambers asserts in a consideration of the importance of staff training: '… in trying to understand projects and to derive practical lessons from them, the staff and their organisation are, if anything, more important than the people they affect. It is the staff who decide policy and execute it'(1969, 8). It is now beyond contention that providing gender awareness training for staff is a fundamental step in any attempt at mainstreaming gender in a given project. Ideally, this kind of staff training should be accomplished long before the staff actually set foot in the field. In the case of the Eagle River project, however, we have not only been forced to deal with a numerically male-dominated staff and the project's patriarchal structure of power relations, but have also faced the added disadvantage of more than nine months' delay before training got underway.

The status of the gender programme and female staff in the project

As can be seen from the discussion of the staffing structure above, women staff (including ourselves) are included in the community-development component, while men control all the technical activities, classified as 'production development'. The simple fact that men outnumber the women lends a greater emphasis to production-related project activities: they are much more highly valued than those within the community development component. Of course one could argue that simply because production-development posts outnumber community-development posts, the former would be more highly valued regardless of sex. However, the fact that most agronomists are men, and most social workers are women, means the two sets of issues are intrinsically linked.

Women staff at field level have expressed their dissatisfaction with the existing gender divisions between the 'technological' and 'social' components of the project, demanding that they too be included in all the more 'technological' training courses offered to staff. One told us: 'we had to force our way in a training course on goat tending, but we work in communities that raise goats' (personal conversation, 1999). Women staff have also questioned the unequal balance of power between the agronomists and social workers, demanding greater participation in the decision-making process as social advisers.

Promoting linking between urban and rural women leaders

In addition to workshops with community members and project staff, we have also run a third series for women community leaders. These aim to sensitise community leaders to the gender programme and grant it greater visibility, while offering leadership training. We have run 13 one-day workshops, some in county towns. Although most participants are rural community leaders, a significant number of urban women were also present. They included women in local government and union leaders. It is important for them to listen to what rural women have to say.

We began these sessions by showing a videotape depicting a 'normal' family day, but one in which gender roles are reversed: while the 'man of the house' cooks, cleans, sews, cares for the children, and realises he is pregnant, the wife goes to work, drinks in a bar with her girlfriends, comes home late, complains about everything, and beats up the husband. Besides being funny and creating a relaxed atmosphere, this tape stimulates discussions on gender relations and women's roles, in which participants can share their experience. The rural women complained of the 'invisibility' of women's work, even when working side by side with their male counterparts in productive activities, caring for the land, planting, or tending to the animals, let alone when, as 'drought widows', they must manage the property on their own.

Conclusion

When we came to the final workshop with women community leaders, we were filled with mixed emotions. It had been a marathon, in which we covered 13 hinterland towns in three months, and many dirt roads in between, reaching 687 women who had responded very warmly to our workshops, and asked us to return. We felt exhilarated with our accomplishment, but we were also afraid that this had been our last trip to the area. Despite the success of our work among the women, there were mounting complaints on the part of local project co-ordinators, who were still accusing us of doing 'feminism' instead of 'gender'. This complaint might cut short the entire gender programme.

Since than, the local co-ordinators have been forced to reconsider our work. Whereas before, workshops with leaders were held in the country towns, the Eagle River project now enjoys increasing support from the local population, including many individuals in local government positions. The field staff report that local residents' visits to the project offices — women in particular — have increased. In the rural areas, there is increased participation of women in project activities geared to production development. In a recent field day for training in the use of alternative animal fodder, for example, 80 of the 200 participants were women. There are reports that in some communities, women are demanding equal participation in decision-making; for instance, when a community has to choose a number of residents to participate in project activities, they insist that at least half of them must be women.

Gender equity cannot be achieved without women's empowerment. This means women's role in rural production cannot be seen as separate from actions which seek to change their status, including within the project's internal structure. We shall not be surprised if colleagues continue to characterise our actions as 'doing feminism' in order to discredit them. It is much more comfortable and safer for them to restrict the aim to 'integrating gender', ignoring its more political objectives. We are busy preparing a new series of gender awareness workshops, this time for public school teachers in the project area, and especially in rural schools. Future plans also include preparing a series of taped programmes to be aired by local radio stations, focusing on health and sanitation, water resources management, education, and sustainable development, all looking at the issue from a gender perspective.

Cecilia Sardenberg, Ana Alice Costa, and Elizete Passos can be reached at NEIM/FFCH, Universidade Federal da Bahia, Estrada de São Lázaro 197, Federaçao 40.210-730, Salvador, Bahia, Brazil.
E-mail: cecisard@ufba.br

Notes

1 This request came as a response to the demands of the International Agricultural Development Fund (FIDA), the international co-operation organisation co-sponsoring the project.
2 Caatinga is common in north-eastern Brazil; it is characterised by small shrubs and trees, including some cactus.
3 The usual age for fifth grade is 10-11, but in rural areas children tend to start school later.
4 Aged five years and older.
5 Agricultural family schools (*Escola Família Agrícola*) were set up by a Catholic priest 20 years ago; these schools offer elementary and high school education for the children of farming families. There are two such schools in the project area.
6 The groups that have been formed have defined the following production interests: two focus on embroidering, two on raising chickens, two on goats, two on pigs, three on garden produce, two on the cultivation of fruits for canning, and two are still undecided.

References

Büttner, Thomas et al. (eds.) (1997) *Hacia una pedagogía de género. Experiencia y conceptos innovativas*, Bonn, Centro de Educación, Ciencia y Documentación (ZED).

Celiberti, Lilian (1997) 'Reflexiones acerca de la perspectiva de género en las experiencias de educación no formal com mujeres', in Büttner et al. (eds.) (1997).

Chambers, Robert (1969) *Settlement schemes in Tropical Africa: A study of organisations and development*, London, Routledge and Kegan Paul.

Costa, Ana Alice and Sardenberg, Cecilia (1994) 'Teoria e Praxis Feministas nas Ciências e na Academia: os núcleos de estudos sobre mulher nas universidades brasileiras' in *Revista Estudos Feministas*, Vol. Especial, pp.387-400.

Women farmers and economic change in northern Ghana

Rachel Naylor

This article argues that representations of rural men and women as victims of structural adjustment measures are simplistic, ignoring the complexities of farmers' engagement with economic change.

The devastation wrought by various aspects of structural adjustment on the lives of the 'rural poor' in many developing countries has been well documented, and the gender-specific impact of adjustment on the 'vulnerable' described (Cornia, Jolly and Stewart 1987). Adjustment has often heralded mass retrenchment for public servants, including those working in agricultural support, an end to subsidies such as those on basic farm inputs, reductions in the availability of credit for farmers, and the introduction of user fees for social services. Observers note that women lose out differently from men as a result of these changes, and many see women as losing out more. This is as true for Ghana as anywhere (Sarris and Shams 1991, Brydon and Legge 1996). Increases in health and education fees are frequently given as an example of changes which hit women hardest, since it is they who are expected to pay for such services for themselves and their children.

There is no doubt that structural adjustment can have negative consequences, especially for the poorest people. But the economic language used in discussions, of 'impact', and 'vulnerability', and even of the 'rural poor', can be simplistic. It can mask the diversity and complexity of rural life and the resourcefulness and power of rural dwellers, particularly women. Rural men and women are actors in a process of continuous change which is played out at community and household level. If we look at how change is negotiated, does adjustment turn out to have straightforward 'impact' of different kinds? Do women always simply lose out? To answer these questions, we need to look at the development process and context in depth.

This article looks at the consequences of the liberalisation of cotton production in communities in Langbensi, an area of northern Ghana, in terms of gender relations in the household and the community. It draws on research carried out over an 18-month period during 1995-6. To understand the processes of agricultural development at village level, I describe gender relations within the household and the wider community, with regards to production and provisioning. I show what has happened to these arrangements when new opportunities resulting from structural adjustment came along. While both men and women seem to benefit, women are taking on new burdens at the same time. Finally, I will indicate some of the lessons for development practice.

Gender relations in north-east Ghana

In 1981, ground-breaking research by Ann Whitehead focused on gender relations and agriculture in Kusasi, northern Ghana. Whitehead's findings demonstrated the critical importance in development planning of looking in detail at the different ways in which women and men are involved in agricultural production, and revealed the danger of making assumptions about the nature of the household, the distribution of resources within it, or the ways in which people interact in the production process. Ann Whitehead's findings have been a major influence on development policy ever since (Gardner and Lewis 1996).

Until then, most development policy had been based on the assumption that the household unit was the basic 'building block' of rural society. The household was assumed to be a unit where production, reproduction, and consumption took place, and one headed by a man. It was also assumed that this unit was characterised by co-operative, altruistic relations, and that its members had a unity of purpose and equality of access to household resources (Mackintosh 1989). Colonial development policy had assisted male household heads with agricultural extension services (Boserup 1989, Elson 1991) because of assumptions about the farm family. Development planners also often made assumptions about gender relations, for example that women did not do heavy work (Rogers 1980). Because of these biases, the majority of women who were involved in agriculture tended to lose out; they could not gain access to new agricultural inputs and techniques of their own, nor did they necessarily share the profits of higher productivity with men in their households.

Ann Whitehead's research in Kusasi gave a very different view of households. She saw the household 'as a site of subordination and domination, of sexual hierarchies of many kinds, and of conflicts of interests between its members, especially between husbands and wives' (Whitehead 1981:92). Kusasi is a polygamous society, where the people within each household live together and share the domestic labour, and work together to produce and consume the millet grown on the household land. The household unit was profoundly undemocratic: the household head controlled it, and different household members had disproportionate access to the resources needed for production, including other people's labour. The vast majority of household heads were men. Household members performed different tasks, and different people provided for others within the household in different ways. Different members also had unequal access to food and to goods produced or held by the household. Access was constrained for women, younger men, and children.

Ann Whitehead also showed that individual agricultural production was of major importance in Kusasi: most crops were produced on the different private farms held by household heads, younger men, and women. But while household heads were able to ask members of their household to work on their private farms, women and younger men did not have the same access to other members' labour. Household heads could also call on members of the community to work their private farms as part of a large communal working party. Women and younger men could gather smaller working parties, but this involved expense. Finally, while individual women had the right to the produce from their private farming, they were in fact obliged to use it to look after their children. In contrast, men were able to use the produce from their private farms for their own consumption.

Households, farming, and provisioning in Mamprusi

My research, undertaken over 15 years after Ann Whitehead's original work, looked at an area to the south of Kusasi: the Langbensi zone of the East Mamprusi district. This area is less densely populated and experiences higher rainfall than the Kusasi area, but like Kusasi is beginning to suffer from severe environmental degradation. It is ethnically mixed, but Mamprusi people form the largest and politically dominant ethnic group. The region forms part of the small Mamprusi traditional kingdom.

Most Mamprusi farmers in the East Mamprusi District are semi-subsistent. The mode of agricultural production is closely tied to marriage, kinship, and the community. Polygamy is the norm in this area; people live in compounds, but these vary greatly in size depending on the stage the household is in the life-cycle[1]. Compound houses consist of one or several households living together which are connected by kinship through the male line. Women move to their husbands' households permanently after several years of marriage. A new household may be formed by a husband, wife, and their children. Longer-established households may consist of a senior man and his wives, with their married sons and their children.

Farms are of small size, and agriculture is labour-intensive. Farmland for the household is passed through the male line to men (unlike moveable property, which is inherited by daughters as well as sons). Individual farmland is lent from the household land to men and women in the household, and to strangers. Lending is usually done on an annual basis, so that it is integrated into the household crop rotation and fallowing plan.

Farming is mixed, which serves to spread risk in an environment where rainfall can be erratic; productivity is very low. A large range of crops is grown, and poultry and animals are reared. The principal crops are cereals (millet, guinea corn, and maize), pulses, (groundnuts, beans, and soya) and vegetables (rosella, other leafy vegetables, and tomatoes). Most of these crops can be sold at market for cash when the need arises, but the crops usually produced for cash are groundnuts, beans, cotton, and tobacco. Many women farmers also engage in an individual capacity in low-return small-scale production and trade. Animal traction in the form of bullock ploughing is used on many plots.

Provisioning

Provisioning is achieved mainly through agriculture. Most men and women work their own private farms. In the Mamprusi area, men are responsible for providing cereal staples for the household's evening meal through organising the cultivation of household land. Culturally, the cereal staples are regarded as the most important food. Because women generally do not farm cereals, they are not seen as 'farmers', and the fact that there are women farmers at all is a fact often strenuously denied by men in the commonly heard phrase, 'women don't farm'. Men who are unable to provide cereals are regarded as hopeless, and women's basic definition of a 'good house' to marry into is one where men provide cereal staples throughout the year. However, women's contributions to daily food are substantial: the household's married women take responsibility for the provision of soup ingredients of this evening meal, taking turns to cook. (Older women with married sons no longer have this evening cooking duty.) Women are also responsible for other food consumed during the rest of the day by themselves and their own younger children, independently of their co-wives. In households that run out of grain during the 'hungry season' before the harvest, women are the only providers for their children, supplying all the foodstuffs.

Gendered access to labour

Hierarchical gender relations in Mamprusi give men a distinct advantage over women here in terms of access to labour and land. Marriage implies that men 'own' women[2] and can draw upon their labour for agriculture. Women are required to plant, weed, and harvest the household cereal farm without payment (as are children). Men work with women and children on these tasks, and also clear land and undertake bullock ploughing. Broken and malformed cereal heads are given to women for their use, but the granaries are controlled by men, who ration out the allocation to the woman whose turn it is to cook. Opposition to this system is not voiced overtly by women, but pilfering from husbands' fields is one form of resistance some women make to these arrangements. As Ann Whitehead found in Kusasi, male household heads who can afford to provide food and entertainment are also able to host large village-wide work parties for weeding work. Women cannot call village-wide work parties.

Access to household labour for women farmers is also problematic, because they have to pay for it with food, and also wait until men's farms have been worked, which may adversely affect their crops. At the same time, women have less time to farm because of their reproductive tasks. A woman farmer described this extra constraint:

'With a woman and a man it's not the same. The man goes to farm. But the woman has to do her work. You do your household work and then go to farm when the sun is hot. Can you work hard then?'

Women tend to have much smaller plots and to farm soup ingredients — 'women's crops' — rather than cereal staples. Women do control the produce from their own plots Older women have the advantage in terms of access to labour where they can draw on their children's labour (after negotiation with their husbands). But even small work parties organised by sons require women to provide something.

As one women's development group leader explained to me:

'If there are weeds at your farm and you have a son, you ask your husband whether your son can go to your plot the next day to weed. If your husband agrees, he will tell the child, 'go and weed your mother's things tomorrow'. Your son can go and organise a small work party so that they weed for you and you provide them with food.'

She added that if the husband refused to release the son or if you had no son, you could call on a junior brother's help if he lived nearby and provide food for his small work party, or you could weed yourself with female friends as a work party, which sometimes might include a co-wife.

Gendered access to land

Except for housing plots in the larger towns, land is not bought and sold in this area of Ghana, and, in practice, the national land registration system does not operate here. The Mamprusi king is said to own all land, but on an everyday level, land is regarded as 'owned' by men farmers who control the household land. Women do not 'own' farms, and must negotiate access to land through men by 'begging' for it. Sometimes women are allocated inferior land. Due to in-migration from further north, increased population and land degradation, the availability of land is reducing, and access to it for women and others who do not own farms is becoming increasingly problematic.

Access to land for the collection of wild bush fruits such as shea nuts, another significant part of women's livelihoods

(Pugansoa and Amuah 1991), is also only possible through men. Women may only gather from husbands' farms (or fathers' farms if they are unmarried).

The effects of liberalisation and adjustment

Cotton had probably been grown for hundreds of years in this area on a small scale, inter-cropped with foodstuffs (Isaacman and Roberts 1995:11-12, Maier 1995:78). From colonial times, there were attempts to introduce it as a commercial crop, but this did not succeed until after Independence in 1957, when the Ghana Cotton Development Board started 'outgrower' schemes[3] with small-scale farmers. Farmers were lent the necessary inputs to grow cotton in their own areas, and the Board bought the produce at the end of the season (Seini 1985). Interviews I held in the Langbensi area revealed that the Board only worked with selected male farmers and, because of its policy on helping these farmers improve general production techniques which included assisted purchase of bullock ploughs, it contributed to the social differentiation between some men farmers and between men and women farmers.[4]

Due mainly to management problems, the Board had all but collapsed by the 1980s, although it maintained a monopoly position. Under structural adjustment, from 1983 onwards, plans were made for the sale of many state-owned enterprises and the liberalisation of many sectors, including cotton. My interviews revealed that as the Board became a private company, entrepreneurs also began to set up their own companies. Other adjustment measures which affected farming in this area at this time included the end of state-subsidised fertiliser and input sales. The government had envisaged that input sales would be taken over by the private sector, but — partly because the market for expensive inputs is perceived to be very small in this area — commercial input supply is weak. As a result, farmers find it very difficult to source manufactured inputs.

Organisation of production in the new companies remains similar to the old scheme[5]: farmers are lent the necessary inputs including seed, fertiliser, insecticide, pump sprays and money to buy-in bullock ploughing or tractor ploughing. They cultivate the cotton on their own land, subject to varying degrees of supervision from the company. At the end of the season, farmers sell their produce to the company, which thus recover the cost of the inputs.

Engaging with change

Men and women farmers in East Mamprusi District have responded to the new opportunity by increasing their cultivation of cotton dramatically. For example, in one village typical of the area, an average of five farmers, exclusively men, were growing cotton throughout the 1980s. In 1990, 13 farmers were engaged in cotton production, which increased to 51 in 1994 and 75 in 1996, including men and women. This is despite the fact that cotton-picking is drudgery, and that men farmers in particular told me that they have fears about the long-term effects of tractor-ploughing and over-fertilising poor soils and spraying insecticide by hand without protective equipment of any kind. As one male farmer observed,

'Fertiliser kills the land. If you use fertiliser this year, next year and the third year, the fourth year if you don't spread fertiliser, you won't get anything at all!'

Some of the reasons for the take-up are obvious, some are less apparent. In a relatively impoverished region, becoming engaged in cotton production provides a source of credit, and then acts as a source of cash income at a vulnerable time of year, since pay-day from the companies occurs at the beginning of the hunger season. Cotton cultivation also helps improve plots, because growing it kills *striga* (a virulent weed) and leaves some fertility in the soil for the next season, when the land can be used to grow cereals. Finally, and top of the list for many farmers, cotton cultivation enables people to get their hands on valuable inputs, such as fertiliser and insecticide, which have not been widely available in the area since adjustment. Some of these inputs can be 'diverted' for use on cereals and beans, to bring production up to subsistence levels, or they can be sold for cash. For some farmers, the process of cultivation has become a subtle performance in image and information management: a delicate game played with company staff. Since companies are keen to attract new farmers, it seems to be a game that farmers can win at present. If one company decides to refuse to work with a farmer because of low cotton productivity in the previous season, she or he may be able to cultivate with another company.

But perhaps the most remarkable thing about the cotton revolution is that up to a fifth of cotton farmers are women. In the areas where I worked and researched, no woman had cultivated cotton for the old Board, because it had only worked with men. But even when the culture of male bias in extension under the old Board lingered on in the late 1980s, when there were fewer private companies and less competition for farmers, women told me that they began cultivating cotton plots by sending sons to register on their behalf (using male names). Soon, women farmers no longer had to be 'invisible' as new companies were founded and had to compete for farmers.

There were other reasons why women were able to take advantage of this opportunity. Women find it easy to negotiate for land for cotton because they only need a particular plot for a year. Cotton is an annual crop and does not require long-term land improvements. Men told me that they are happy for women to cultivate the crop because it is of value in the crop rotation system, adding fertility and decimating *striga*. As one woman farmer noted, 'the man will get a good crop next year [when he cultivates on the area planted with cotton this year] because of the fertiliser you spread'.

Women told me that they operate their plots on very small risk margins and that they find it more difficult than men to obtain both credit and access to labour. The provision of credit by the cotton companies is a major reason why women are taking up cotton cultivation. Women also reported that access to credit to buy in ploughing helps them to negotiate for ploughing to be done on time. My research showed that some companies have also provided credit to farmers to pay for weeding labour, which has been particularly helpful for women.

Difficulties faced by women who grow cash crops

Many women take up the opportunities in cotton production, but they report that they still experience problems related to their negotiating position over labour and their own time. One woman farmer described her difficulties in getting her cotton planting done on time:

'The tractor came early [to plough the cotton plot] but we hadn't finished planting [our husband's] millet. He 'owns' us. You eat from his mouth. If you eat from his mouth, can you eat from your own mouth? ... If we finish all his planting, the millet, the maize and groundnuts, absolutely all, then we wives can start our own planting.'

Access to labour for the cotton harvest has also proved an enduring problem for women:

it coincides with the harvest of the main crop, millet, which they are required to help with. Older women with children are more likely to be able to cultivate cotton, because they can draw on their labour. But women's cotton is more likely to be harvested late and is prone to being destroyed, eaten by cattle which are allowed to roam in the bush farms after the millet harvest.

In dealing with cotton company staff and their (male) farmer representatives in the village, women also report that they are losing out in the distribution of chemicals and fertilisers. This is because they are reticent to ask and press for their allocation of inputs, which requires a certain assertiveness. Women also still prefer to send sons to cotton company meetings, where farmers are asked to share their experiences in planning for a new season's cultivation. As a result, they lack input into decision-making. Often, women reported, they are not invited to these meetings anyway.

Looking at the benefits that women and men gain from cotton, profits are generally low but women are more likely to spend the cash on food, whereas men are more likely to spend it on consumer goods such as cloth, lanterns, bicycles, and zinc roofing. This relates to the existing gender role, which expects women to be the providers of last resort for their children.

One successful male farmer described how cotton had benefited him over the past two years:

'Last year I bought roofing sheets. You can't grow maize and sell it to buy roofing sheets, you will eat it. This year I bought the wood and nails for the roofing and a cow.'

The traditional women's leader in one village explained how and why most women spend their cotton profit:

'You can use it to buy food if your husband's food finishes so that you and your children can eat. If the time for school fees has come and your husband has no strength, you can sell part of the food you bought and use it to pay for the school fee.'

Alternatives: meeting strategic needs

To ensure a better future for all, men and women farmers need to co-operate within the household, albeit a changed one. Bina Agarwal has shown (1994), in a very different context, the importance of equality of access to the means of production in empowering women. Without negotiating rights to land, labour and over decision-making, it is difficult for women to do more than satisfy basic needs by taking on new responsibilities. It is clear that cotton is helping women to shoulder the increasing burden of meeting their children's basic needs, rather than to raise living standards in any other sense, or to make a strategic change to their condition.

It is by working with other agencies on the scene, such as Langbensi Agricultural Station, a non-government organisation (NGO), that women have begun to address what Caroline Moser (1989) terms 'strategic' gender needs. The Station was set up in the early 1970s by the Presbyterian Church of Ghana to provide extension services and sell inputs. It currently works with over 60 men's, women's and mixed groups in the area to improve rural livelihoods and facilitate farmer-led research (Kolbilla and Wellard 1993).

By working with and drawing on the expertise and the lending power of the Station, the women's groups have increased not only their material status but also their capacities to organise and their levels of confidence. Groups tend not only to work together on group farms but also to help each other on their personal plots, easing labour access problems. Women reported to me that menfolk find it difficult to refuse when they ask for time off from household farming to attend to the group plot.

Previously reticent in the company of men, women's groups have begun to negotiate permanent access to group plots for tree-planting and farming and to challenge encroachments on their land. This enables them to make long-term improvements to the land and to grow trees as a long-term investment in timber and in the environment, as well as easing the reproductive chore of firewood collection.

In one village where I undertook research in April 1995, the women's group's agroforestry and farm plots were encroached upon by a male farmer from the local town. Many of the tree seedlings were destroyed. The problem was reported to the village chief, but he decided to appease the town farmer and take no action. Although 'affairs of the world' are still felt to be a man's domain, the women's group agreed that the issue could not be left at that. They sought a meeting with the chief and elders (without the assistance of Station staff). The women demanded financial compensation for the loss of the trees and the immediate allocation of a new plot for farming, as it was already late in the planting season. All the women's demands were met, and the town farmer was charged to leave the women's land for them at the end of the season.

Groups of women farmers now sit with men's groups in meetings facilitated by the Station in the villages. At these meetings, men and women have begun to discuss and find ways of exploring the difficulty that the farm family faces. Even their particular difficulties in the household as women, in terms of the burdens that they shoulder and the problem of access to labour, are beginning to be aired. A senior manager at the Station commented,

'At first, men and women couldn't sit together at meetings. Now, not only can they sit together, but women can also speak out. Men are beginning to see that the women's contribution is important. They are able to realise that they are partners. Of course, in private, the men will always admit that they can't do without the women, but not in public.'

Conclusion

These slow, empowering changes also have the potential to increase the ability of women in communities to engage with, and profit from, the many 'external' agricultural agencies they interact with in the development process — many of whom continue to show biases based on misunderstandings of household dynamics and the agricultural system.

In one remote agricultural community in Ghana, on a hot and dusty day in June, I witnessed representatives of an NGO who had brought a photographer along to take pictures of a group of women farmers. The group was planning to process shea nuts into shea butter for sale, in order to boost their incomes. The women sought funding from the NGO for the initial capital, and the NGO would use the photograph to support the application to a Northern donor organisation. The group wanted to show the donors how they would process the shea nuts in a way that would convince the donor to support the project; they therefore posed with mortars and pestles. But the scene was not as straightforward as first meets the eye: the mud and thatch building that formed the background of the photograph was in fact a grinding mill where the women, once they had enough cash, would undertake the initial processing, foregoing the need for the hand tools.

The women's 'image management' here illustrates the idea that change is a negotiated and manipulated process, played out in the development encounter. The dominant development discourse may be one backed by international financial institutions and multilateral donors, but the large agencies do not always dictate who has the power to privilege their discourse in any particular situation. Where there is leeway for action, people on all sides may use the power or 'agency'[6] they have to improve development outcomes for themselves. By focusing on our need to understand this process, we can learn how better to support people in their struggle for equitable development, and in their negotiations for positive outcomes from change such as economic adjustment.

Rachel Naylor is an Action Researcher with the Rural Development Council for Northern Ireland. She has carried out research for Oxfam GB, including work on peace-building (van der Linde and Naylor, 1996), a country profile of Ghana (forthcoming), and various gender and development assignments. The fieldwork for this article was carried out for a research degree in rural development at the Department of Sociology and Anthropology, Hull University.

Notes

1. This term, developed by Goody (1958), refers to the way in which a new household may begin as a small unit which has split off from a larger household. It may consist of a young married man and woman and their children. Over time it will expand to become a large grouping, perhaps including married children and grandchildren. New households may then split off from this unit, beginning the cycle again.
2. *N-suu*, to own, implies control over things and people.
3. A system whereby an agency organises and controls much of the production process, such as input supply and cultivation practices, but contracts farmers to produce the crop on their own land.
4. Devereux (1992:33-4) also noted the increased socio-economic differentiation which resulted from the introduction of bullock ploughing technology in Kusasi.
5. However, I discovered that while the organisation is similar, the culture of the companies appears to have changed. For example, in commercial concerns, staff work to bonuses. The companies aim to expand, and have to compete for farmers.
6. This is the idea that people at the 'grass-roots' actively engage in shaping their own situations rather than responding to activity at the 'centre', developed in a critique of dependency theory (Long and Long 1992).

References

Agarwal, B (1994) *A Field of One's Own: Gender and Land Rights in South Asia*, Cambridge University Press.

Boserup, E (1989) *Women's Role in Economic Development*, Earthscan, London (first published 1970).

Brydon, L and Legge, K (1986) *Adjusting Society: The World Bank, the IMF and Ghana*, Tauris, London.

Cornia, G A, Jolly, R and Stewart, F (1987) *Adjustment with a Human Face*, Clarendon Press, Oxford.

Devereux, S (1992) 'Household responses to food insecurity in northern Ghana', DPhil thesis, University of Oxford.

Drucker-Brown, S (1993) 'Mamprusi Witchcraft, Subversion and Changing Gender Relations', in *Africa*, 63(4):531-549.

Elson, D (1991) *Male Bias in the Development Process: An Overview*, Manchester University Press.

Gardner, K and Lewis, D (1996) *Anthropology, Development and the Post-Modern Challenge*, Pluto, London.

Goody, J R (1958) *The Development Cycle in Domestic Groups*, Cambridge University Press.

Isaacman, A and Roberts, R (1995) *Cotton, Colonialism and Social History in Sub-Saharan Africa*, Heinemann, Portsmouth NH.

Kolbilla, D and Wellard, K (1993) 'Langbensi Agricultural Station: Experiences in Agricultural Research' in K Wellard and J G Copestake (eds.) *Non-Governmental*

Organizations and the State in Africa: Rethinking Roles in Sustainable Agricultural Development, Routledge, London.

Leach, M (1992) 'Women's Crops in Women's Spaces', in E Croll and D Parkin (eds.) *Bush Base: Forest Farm*, Routledge, London.

van der Linde, A, and Naylor, R (1999) *Building Sustainable Peace: Conflict, Conciliation and Civil Society in Northern Ghana*, Oxfam, Oxford.

Long, N and Long, A (1992) *Battlefields of Knowledge: The Interlocking of Theory and Practice in Social Research and Development*, Routledge, London.

Mackintosh, M (1989) *Gender, Class and Rural Transition: Agribusiness and the Food Crisis in Senegal*, ZED, London.

Maier, D J E (1995) 'Persistence of Precolonial Patterns of Production: Cotton in German Togoland 1800-1914', in A Isaacman and R Roberts (eds.) *Cotton, Colonialism and Social History in Sub-Saharan Africa*, Heinemann, Portsmouth NH.

Moser, C (1989) 'Gender Planning in the Third World: Meeting Practical and Strategic Gender Needs', in *World Development* 17(11):1799-825.

Pugansoa, B and Amuah, D (1991) 'Resources for women: A case study of the Oxfam sheanut loan scheme', in T Wallace and C March (eds.) *Changing Perceptions*, Oxfam, Oxford.

Rogers, B (1980) *The Domestication of Women: Discrimination in Developing Societies*, Kogan Page, London.

Sarris, A and Shams, H (1991) *Ghana under Structural Adjustment: The Impact on Agriculture and the Rural Poor*, New York University Press/IFAD, New York.

Scott, J C (1985) *Weapons of the Weak: Everyday Forms of Peasant Resistance*, Yale University Press, New Haven.

Seini, W (1985) 'The economic analysis of the responsiveness of peasant cotton farmers to the price incentives', in *Ghana: Technical Publications Series 51*, ISSER, University of Ghana, Legon.

Whitehead, A (1981) '"I'm hungry mum": The politics of domestic budgeting', in K Young, C Wolkowitz and R McCullagh (eds.) *Of Marriage and the Market*, CSE, London.

'Lazy men', time-use, and rural development in Zambia

Ann Whitehead

This paper examines how we measure work and labour in agriculture in rural sub-Saharan Africa.[1] It has increasingly been recognised that many rural African women have heavy workloads; in some accounts, this is contrasted with apparently light work burdens for men. I argue that in making women's work visible, where once it was not, it is possible to slip into thinking of African rural men as not doing very much at all. There is a danger in some policy discussions of producing an image of rural men as standing idly by, while their wives and daughters are overburdened with work.

Discourses of the inactive or 'lazy' man in Africa have deep historical roots, and have been an central component of racist attitudes towards Africans. This paper presents a historical example of colonial discourses of the 'lazy' African (the Lamba in Zambia), and examines a piece of modern research on gender and labour: a time-use survey conducted in rural Zambia (Allen 1988). This piece of research has been influential in discussions of gender relations in rural Zambia; it has, for example, been taken up by a recent World Bank study on gender and poverty in Africa (Blackden and Bhanu 1999). The image of the 'lazy African man' produced in the historical accounts is apparently proved in the contemporary time-use survey, but I argue here that the latter is misleading, and that assessments of work and labour are complex and often subjective in both accounts.

Over the past 15 years, time-use studies have been employed by development planners to design programmes based on an understanding of the social divisions of labour between women and men, in particular contexts, and of differences in time-use. I suggest in this article that time-use studies embody value judgements about what constitutes work, and about how researchers and planners categorise this. In using these kinds of tools, greater attention must be given to how different kinds of work are understood, to the relationships between them, and the context and location in which they are done.

The 'lazy African' in colonial discourses

The multiple projects of British and French imperialism and colonialism in sub-Saharan Africa were based upon a kaleidoscope of ideas about Africans and African society. At the core of the ways in which Africans were thought, written, and spoken about was the constructed image of Africans as 'different'. As recent historians of colonialism have stressed, these ideas were constantly changing, challenged by the unforeseen and infinitely varied reactions of Africans themselves to the colonial experience (Stoler and Cooper 1997). Economic practices and economic relations were a fertile ground for European ideas about Africa. Colonialism transformed African economies, for example by creating new fields of employment for unskilled porters and labourers, and later for low-grade public service employees. At the same time, far-reaching processes of economic transformation began as African land was mined for minerals on an industrial scale, or put to the monoculture of export crops after being sold to European settlers, and as huge new internal markets for agricultural and non-agricultural goods opened up.

As new relationships were forged as a result of economic change, and new economic practices came into being, ideas about land, property and exchange, about labour and work, and even about time, personhood and identity were transformed and recreated. Documents from throughout the colonial period are replete with ideas and value judgements about Africans' presumed values regarding work and labour. Although there are some realistic assessments about work and incentives in the changing local economies, most Europeans expressed negative stereotypes about 'the' African's antipathy towards, and limited capacity for, work and labour.

Creating the stereotype of the 'wild' and 'lazy' Lamba[2]

I have chosen one case study to illustrate the relationship between these stereotypes held by many Europeans, particular aspects of the colonial project, and the social relations brought about by colonialism.

The mines of the Zambian Copperbelt were developed on land which at the turn of the century was occupied by Lamba people, as well as by Swahili traders and slave raiders.[3] Lamba fortunes were at very low ebb in the 1890s when the British, in the form of the British South Africa Company under Rhodes, established control in the Northern Rhodesian part of the Copperbelt. Suffering from food shortages (an effect of Swahili slave raids) and small-pox (which had been brought to the area by one of the early British South Africa company expeditions in 1890), the Lamba people initially sought to avoid European tax and labour demands, and acquired a reputation for independence and indolence.

During the first quarter of the 20th century, ever-larger tracts of land were appropriated for European mining and farming interests. The Native Reserves Land Commission began formalising these land appropriations in 1926, and as a result, many more Lamba people were moved from their land to Native Reserves further away from the growing urban centres of the Copperbelt, and restricted to about a fifth of their former area. 'To this day, memories of this loss of land and its attendant hardships are the principal focuses of the Southern Lamba's internalised sense of grievance and resentment.' (Seigel 1989, 355). In part, the forced removals were designed to induce the Lamba into farm or mine labour.

However, the proximity of the Lamba to the mines and the growing urban areas enabled them to market grain and vegetables. Thus, Lamba could earn cash incomes without being employed on either the settler farms or the European mines, both of which were avoided by the Lamba, because of their poor pay and conditions (ibid.).

In 1989, Seigel wrote that 'Europeans have described the Zambian Lamba in remarkably similar derogatory terms for the past 80 years or more … as 'timid, lazy' and 'backward…' (Seigel 1989, 353). His evidence comes from the 1926 Native Reserves Commission, which was taking submissions from local interested parties in order to establish Native reserves and allocate the land thus alienated to various European interests. The submissions contain many stigmatising stereotypes about 'lazy' or 'backward' and 'stupid'

Lamba, made by Europeans who were frustrated by the Lamba's determined resistance to being forced into agricultural labour and mining.

Numerous other sources suggest that throughout colonial Africa, those European settlers and farmers who most needed African labour were most likely to berate natives as idle and lazy, since Africans preferred to work on their own farms within less alienated forms of labour.[4] Yet some colonial observers understood perfectly well the economic rationality of peasant families. For example, talking of the Lamba at about the same time as the 1926 Native Reserve Commission evidence, a sub-district colonial officer wrote in his annual report:

'The local (Lamba) native is not very popular with a number of employers of labour, who consider him particularly stupid and dislike his tendency to work only for a month or two at a time. Agricultural products have a ready market in an area with so much mining activity and the local man naturally prefers to get his money in ways by which he can live at home.'
(Zambia National Archives, quoted in Seigel 1989, 356)

While Lamba men continued to stay out of the mines, they did not prosper as agricultural producers, becoming steadily impoverished and by all accounts very demoralised (Seigel 1989). High population density on the Lamba land reserves limited their farming and farming income. After a devastating famine in 1940/41, the colonial administration introduced a resettlement scheme, but this was such a resounding failure that in 1953 African peasant farming families were brought into the area from Southern Rhodesia to show the 'apathetic' Lamba, 'by demonstration the possibility of advancement in agriculture.' (Northern Rhodesia, African Affairs Annual Report for 1952, quoted in Seigel 1989, 359)

The stereotyping of the Lamba as indolent and apathetic by British mine managers, farmers, agricultural officers, and missionaries has its roots in the Lamba people's unwillingness to undertake European forms of labour and agriculture and their ability to keep aloof from colonial labour relations. Lamba rural producers also resisted the colonial schemes for agriculture that characterised the rural economic development of the 1950s. We have little access to information on what Lamba men and women farmers thought of this scheme, but according to Seigel, resettlement was compulsory — harking back to the earlier resettlements which had been so disastrous for them. The cultivation requirements of the colonial agricultural schemes were 'onerous' (Seigel 1989:359), suggesting that there was little room for farming families to mix cash and food crop production to their own satisfaction.

Derogatory stereotyped views about the Lamba were not only held by Europeans: '... these stigmatising Lamba stereotypes have not been confined to Europeans alone. African townsfolk on both side of the Copperbelt have shared essentially similar invidious stereotypes about the Lamba for at least the last 40 years.' (Seigel 1989:353) The well-paid skilled workers on the mines, many of whom had lived in Copperbelt towns for years with their families, looked down on impoverished local 'country bumpkins' and poured scorn on the dress, the food, and even on the morals of the Lamba. Although they too had been rural producers in their own home areas, they had taken up mine work from early on and had developed very different attitudes towards work, and to ways of earning a living. From the negative stereotyping of Lamba by other ethnic groups on the Copperbelt we can see the ways in which different ideas about economic value, labour and work, quality of life, and ways of life are in play in various areas of the Copperbelt.

Defining work: Feminist economics and time-use audits

This section examines the ways in which current approaches to work and labour in sub-Saharan Africa embody value judgements which are similar to those discussed in relation to the Lamba under colonial rule, leading to distorted accounts of the division of labour between women and men.

Many current accounts of men's and women's work in sub-Saharan Africa incorporate ideas of the ways in which the division of labour between women and men is an area of contestation and negotiation. This awareness owes much to the work of feminist researchers and activists: second-wave feminists[5] have persistently argued that women's unpaid work within the home must be recognised as an activity which has economic value, although they have acknowledged that it is not done under the same conditions as waged work. Women produce many goods and services within the home without which family members would be unable to enter the labour force. In 1995, the importance of measuring women's unpaid work was adopted as a major outcome of the Beijing conference (Beijing Platform of Action 1995).

However, it is difficult to assess adequately the precise nature of the economic contribution of men and women in terms of productive and reproductive work, since the categories for collecting data on work and production used by national accounting systems centre on activities which have a visible market value. Unpaid domestic work is not the only economic activity rendered opaque by these categories. Self-provisioning (subsistence) production, especially in agriculture, is also notoriously poorly captured by conventional data collection.

The major methodological tool that has been developed to estimate unpaid work is time-use auditing, which classifies activities into 'work' and 'non-work'. The United Nations Statistics Division defines 'non-work', as time spent in 'personal care and free time', which 'includes bathing, sleeping, eating, time related to personal medical attention, resting, organisational participation, sports and games, socialising and media related activities (reading, television)' (United Nations Statistics Division[6]). All other activities are classified as 'work'. This classification attempts to make visible the economic value of unpaid work in society, and to capture the relative work burdens of men and women, but accurate time-use data are extremely scarce, and far from simple to collect and interpret. Moreover, time-use studies, which provide an apparently objective set of measurements about gender equity, often embody value judgements about activities. These judgements colour the classifications of activity being used. For example, one problem is judging which activities are classified as work or leisure. How do we distinguish between work and non-work? This is a particular issue when considering how to categorise child-care, which is usually treated as household maintenance or reproductive work. But is every minute of child-care work? Is none of it refreshing leisure? Does the bedtime cuddle which gives you the strength to face another hour at the computer, or to do the laundry, count only as a burden? Can we distinguish child-care that is work from child-care that is pleasure? Should we? If so, when and how? Child-care often overlaps with other activities, and this is may be classified as 'multi-tasking' (taking on two or more work activities simultaneously).

Seemingly technical efforts to assess work-loads embody values about what is work, and about how to measure overlap and multi-tasking. It is not possible to avoid this problem by asking women and men themselves to classify their own activities as burdens or pleasures. Their views, like those of the researchers, will be filtered through culture and expectations, which have often routinely screened out some of the work which women do. Researchers working with them therefore have to consider these issues and take a stand on them. This is a particular problem when activities are being classified and measured across cultural divides, where a high proportion of the goods and services produced and consumed in a household may be produced by processes that are not governed primarily by market principles. Time-allocation data in these circumstances has to be approached with extreme care, especially when considering the differences between the work burdens of women and men.

An example from Zambia

Very few national household time-use studies for rural African households exist, although time-use data is of particular relevance because the majority of rural households produce at least some of their own food. The domestic nature of this food production means that it does not necessarily result in statistics on rural hours of employment and labour-use. Some of the best available data are from case studies from rural Zambia, which come from farming systems surveys carried out by the nation-wide Adaptive Planning Research teams (APRT). The context for these surveys was the Zambian Government's interest in tracking the progress of their nationwide, tightly controlled package for the introduction of smallholder hybrid maize as a solution to the problem of national food security. Over time, the APRTs have published a good deal of data which it has gathered from various rural centres[7]. These studies are an important source of data on men and women's time inputs to a range of economic activities, and contain some of the only data that exists on children's work in rural sub-Saharan Africa; however, they have to be interpreted with care.

The research teams collected raw data according to a common set of classifications. (see Table 1) Activities are divided into agriculture, food preparation and household activities, building, foraging, business, working for others, leisure, school, ill health, and meetings. Time not accounted for in these activities is classified as 'resting'. APRTs also use a common set of socio-economic categories for households, dividing them according to the amounts of bags of crops produced. Some of the time-use data show a pronounced imbalance in women's and men's workloads. In some rural areas, women appear to be busy all the time, while men's main activities seem to be resting, visiting friends, and 'leisure'. These kind of time-allocation findings have influenced much of the literature on Zambian gender relations.

An example is Allen (1988), which draws on a detailed examination of time-use data from Mabumba, an area in Luapula, which was collected during the author's employment as a social scientist with the ARPT. Allen comments: 'The extent of men's inactivity is truly astonishing. Male humans in Mabumba are basically dependent on females for their food and succour from the cradle to the grave.' (Allen 1988:43). This judgement rests on the evidence of time-allocation information from just 13 households, including male- and female-headed households, which have varying levels of economic activity. The Mabumba research shows an average woman to be spending 43 per cent of available time[8] in productive work; the figure for the average man is 12-13 per cent. Over the year, this averages as a six-hour working day for an adult woman, compared to less than two hours of work per day for a man. Two pictures of men emerge: first as idle wastrels, pursuing selfish interests such as socialising with other men and drinking; and second as child-like dependents. This second picture gives rise to the title of Allen's paper:

Activities in order of household importance	Resting	Leisure	Household food	Agriculture	School	Ill health
Female time allocations (hours per year)	1,922	135	1,031	657	27	336
% of total time	43.9	3.1	23.5	15	0.6	7.7
Ranking	**1**	**5**	**2**	**3**	**10**	**4**
Male time allocations (hours per year)	2,463	847	8	243	243	156
% of total time	56.2	19.3	0.2	5.5	5.6	3.6
Ranking	**1**	**2**	**12**	**4**	**3**	**5**

Table 1: Comparative involvement of male and female adults in major classes of activity in Mabumba, Luapula Province 1986/87. Source: Allen 1988. (Continued opposite)

'Dependent males: The unequal division of labour in Mabumba households'.

Allen's study is a relatively obscure publication, yet it is frequently cited in discussions about gender relations in Zambia, and it has been the starting point for a number of evaluations of intra-household gender equity in Zambia.

A particularly important document in this regard is the 'Status Report on Poverty' by the World Bank's 1998 Special Programme for Africa (SPA), whose principal theme is gender (Blackden and Bhanu 1999). This presents an analysis of the gender division of labour and the relative contributions of men and women to national economies which owes much to the insights of feminist economists. It emphasises the longer hours that women spend in productive work and in agriculture compared to men in sub-Saharan Africa. Its account of what African time-use studies reveal about the relative work burdens of men and women takes Allen's findings as its starting point.

The high profile and potential influence of this World Bank document make it particularly important to re-examine Allen's study, on which so much of the argument rests. The tables included here reproduce data from Allen (1988), which show how he has aggregated the initial data (Table 1), and alternative tables in which I have reworked the raw data, using slightly different classifications (Tables 2 and 3). I also include some illustrative data from Zambia's Northern Province from another study which uses the APRT data (Evans and Young 1988), Table 4.

Who is doing the work?

Allen's original interpretation (Table 1) suggests that women in the Luapula community are engaged in much more productive and reproductive activity than men, and that they carry a very heavy burden of farming work. However, a comparison of Tables 1-3 shows that Allen has chosen a number of ways of aggregating the data which magnify the differences in workloads between men and women. For example, the averages for women's work and men's work do not distinguish between different kinds of men and women. The average conceals the different workload of female household heads, wives, and other adult female family members, such as daughters, or schoolgirls over 15 or mothers, and there is no information that makes it possible to reconstruct their relative contributions to the 'average' woman's load. Similarly, we do not know how the work burdens of older or younger men (including those in school) compare with the 'average' man's load.

Activities in order of household importance	Business	Meeting	Paid work	Work for others	House building	Foraging	Total activity
Female time allocations (hours per year)	66	89	47	49	5	16	2,458
% of total time	1.5	2	1.1	1.1	0.1	0.4	56.1
Ranking	7	6	8=	8=	12	11	
Male time allocations (hours per year)	104	110	93	60	30	22	1,917
% of total time	2.4	2.5	2.1	1.4	0.7	0.5	43.8
Ranking	7	6	8	9	10	11	

Production and household maintenance

Allen's data classifies the work of household maintenance together with other productive work. In all, his 'productive work' category includes farming, wage employment, and business, together with household maintenance. Allen's decision to include domestic work as 'productive' activity falls in with the recommended approach from feminist economics, although it is a very unusual decision, especially for 1988.

In contrast, in my own reworked tables (2 and 3), I have divided this productive work category into three: farming, other economic activity (including foraging, working for others, employment, and business), and household work. Reworking Allen's data under these categories confirms his account of the time-consuming nature of household maintenance work for women in Mabumba. The average time spent on this work per woman is 1,056 hours per year, while a man does only 38 hours. The time spent in such work is about 10 per cent lower for women in female-headed households. In Table 4, which relates to the Northern province, wives' annual domestic work hours appear very similar to those from Mabumba, and other time-allocation studies across Zambia have similar amounts of time spent on domestic work. It is clear that the daily task of feeding and meeting basic needs within

Table 2: Recalculated time allocations of men and women by household headship*

	All households (13)		Male-headed households (8)		Female-headed households (5)	
	Women	Men	Women	Men	Women	Men
Average age (years)	34	34	33	38	38	23
Total annual hours spent farming	657	243	679	267	594	175
Other economic activities (hours)	178	279	123	282	308	269
Subtotal 'productive' work (hours)	835	522	802	549	902	434
Total annual work hours	1,891	560	1,887	579	1,859	496

* adapted from Allen (1988), tables (ii), (iii), (iv) and 3.1

the household in rural Zambia is culturally assigned to women, and constitutes a major area of their energy expenditure.

Men's lighter work burdens in Allen's account are thus partly explained by the unequal responsibilities for domestic labour. This is not an outcome of peculiar features of gender relations in Zambia. African time-use studies all report similar findings of high levels of time allocated to domestic or reproductive work. Women's time spent in this way in sub-Saharan African households seems to average about four hours a day. Although men do some household maintenance in some contexts, especially house repairs, the average time spent can be very small — certainly less than one hour a day[9]. Significantly, while collecting water and fuel are very demanding for some African women in certain areas, and in certain seasons, preparing food before cooking, and cooking itself, are the main demands on women's time. These are also the tasks which men are least likely to take on. It looks as if women's reproductive and domestic burden is non-negotiable. Food processing and preparation, cooking, and collection of fuel and water are all essential to the household's survival, as is producing items such as cooking utensils, sleeping mats, and soap. The high proportion of women's time spent in such activities has implications for women's capacity to do other work.

The same phenomenon occurs beyond sub-Saharan Africa, not just in the developing world, but in many developed countries. UN data on global time allocations show imbalances in time spent doing unpaid work between men and women throughout the world; women therefore spend a smaller amount of time resting and taking part in leisure activities. However, in Europe, North America or Australia — despite the fact that women carry heavy burdens of household maintenance work (and this remains an important focus for feminist politics) — dominant policy debates on men's time-use rarely focus on the inequality of this division of labour. Provided a man is doing paid work, he does not tend to be castigated as a dependent or as idle in public policy documents if he does no housework.

Why should African men be treated differently with regard to their avoidance of domestic tasks? In Allen's case this allows him to emphasise the imbalance between men and women in their inputs to production, and to paint a picture of rural production as a primarily female domain. This is the way in which his findings have been used in the World Bank's 1998 SPA Poverty Report.

*Table 3: Recalculated time allocations of men and women by socio-economic category**

Socio-economic category	0-5 bags		6-19 bags		20+ bags	
Household headship	Female (2)	Male (1)	Female (2)	Male (3)	Female (1)	Male (4)
Time allocations (annual hours)	Women	Men	Women	Men	Women	Men
Farming	794	198	190	202	538	184
Other economic activities*	190	432	417	507	105	335
Subtotal 'productive' work (hours)	984	400	728	616	622	842
Household work	1,111	4	983	15	920	9
Total annual work hours	2,095	404	1,711	631	1,542	851
School	–	267	–	329	166	–

* includes foraging, business, employment, work for others

Table 4: *Recalculated time allocations of husbands and wives by socio-economic category in Chunga, Northern Province 1982/83. Adapted from Evans and Young (1988). Chunga is a patrilineal community.*

Socio-economic category[1]	A		B		D		E		F	
Time allocations (hours per year)	Women	Men	Women	Men	Women	Men	Women	Men	Women	Men
Farming	467.5	179	202	152	430	438	456.5	350.5	513	354.5
Other economic activities[2]	135.5	808	139	316	141.5	498.5	159	1,131	89[3]	303
Subtotal 'productive' work	**603**	**987**	**241**	**468**	**511.5**	**936.5**	**615.5**	**1,481.5**	**602**	**657.5**
Household work	884	82	817.5	78	1,131	83.5	1,142	77.5	1,115	66
Total annual work	**1487.5**	**1,069**	**1,958.5**	**546**	**1,642.5**	**1,020**	**1,757.5**	**1,558**	**1,717**	**723.5**

1 ABCDEF refer to the following categories: A are households producing virtually no maize for sale, F are those producing most hybrid maize sold in large quantities. No household was classified as category C.
2 Includes foraging, business, employment, work for others.
3 Artificially low figure as data are missing for Oct/Nov.

Seeing agriculture as part of a livelihoods strategy

Allen's analysis of time-use also fails to link men's and women's agricultural activities to their other income-generating activities. The farm is not set within the context of domestic production as a whole. Allen records the overall ratio of women's to men's time input to family farming in Mabumba as 2.75 to 1. It is clear from this that farming depends heavily on women's labour input.

By concentrating on tracking women's and men's time inputs to agricultural activities on the family farm, Allen's data analysis was following the conventions of much 1980s research into farming systems and agricultural economics, which paid little attention to off-farm activities. However, this seriously misunderstands the economic strategies of rural households. In the 1990s, approaches to rural livelihoods have stressed that rural households try not to rely solely on their own subsistence production, or on agricultural income, partly because of climatic and market risks (Reardon 1997). Although having many income sources gives poor returns, combining a few may be more lucrative. Diversification into non-farm work is not just a survival strategy — it can be a source of capital for farming.

If one turns to the Mabumba data and considers the time women and men spend on all economic activities (excluding household maintenance work) the ratio of female to male time input becomes 1.6 to 1, because men are more occupied than women in employment and business activities outside the family farm. In Zambia, the sources of non-farm income are many and various. They can include income from migrant labour, from micro-enterprises, from trading, and from local casual farm labour. In Table 4, focusing on Zambia's Northern Province, Evans and Young draw on APRT data to show that the hours spent by men on off-farm activities can be high. In poorer households, men tend to be doing casual wage work, but in better-off households men are engaged in what APRT terms 'business' activities. In Table 4, households producing large quantities of hybrid maize record very high numbers of male work hours spent off farms — probably in 'business'. There is almost certainly a link between the success in farming and the profits from their business enterprises.

A narrow focus on labour use in farming, as in Allen's account of the female nature of Mabumba farming, misses important linkages between on-farm and off-farm activities, including the role of off-farm activities in ensuring agricultural success, for example, in providing cash for farming. This implies that men's non-farm economic activities are not connected to the livelihood of the household (or indeed to family welfare), and renders rural men's contribution to farming and the wider economy invisible. It also results in an exaggerated picture of inequality in the relative inputs of men and women to the success of the farming enterprise.

The socio-economic context

Perhaps the biggest problem with Allen's analysis is the paucity of information about the 13 households in the Mabumba study, and about the wider economic and social context of the area. To make sense of the bald time-use data he gives us, we need to know much more about the context. For example, without demographic information about the social composition of households, it is difficult to interpret the findings relating to female-headed households. Two-thirds of the group of households which produces little for the market, and in which women do more economic work than men, are female-headed (Table 3); to understand the dynamics of this behaviour we need to know how many of the male members of female-headed households are young. Are they at school, or are they unproductive for any other reason?

Allen's study also provides little background information about the nature of socio-economic stratification between households, and the different economic opportunities and kinds of livelihood strategy open to households in Mabumba. We have no idea what kinds of constraints exist, for example on households doing different kinds of farming. Are some households too poor to farm, in that they lack the resources required to raise their production? What are the local labour markets like for men and women? This set of issues is particularly important because Allen's findings are rather out of line with the other Zambian time-allocation data. In many areas where hybrid maize production was successful in the 1980s, men's labour input to farming equalled, or was in excess of, that of women (see, for example, Skonsberg 1989, Kumar 1994, on the Eastern Province). This pattern is apparent in Table 4.

It is also important to possess information on the socio-economic context, including the local labour markets open to men and women, in order to understand Allen's category of 'resting'. Harrison (1999) demonstrates that 'resting' as used in the APRT studies is essentially a residual category, into which all time that is otherwise unaccounted for is placed. The studies provide little information on what actually takes place during these 'resting' periods. Harrison's research, undertaken in a location very close to the site of Allen's study, suggests that far from being idle, men are occupied in various activities such as developing social networks, making contacts, gathering information, and attempting to find work and business opportunities. These tasks are essential to generating income in areas where opportunities are scarce.

A quite different kind of contextual information is needed on the local experience of labour migration. This has potential relevance to women's and men's decisions about their use of time. On this, too, Allen's short paper is silent. In the late 1980s, Zambia was one of the most highly urbanised countries in sub-Saharan Africa. For 50 years, many men born in rural areas had worked in the mining towns of the Copperbelt. Female migration began later, and never reached the same rates as that of men, although it has long been more common than has been supposed (Ferguson 1990). In the late 1980s, workers returned to the countryside in large numbers as a result of the down-turn in the mining industry and the urban economy. We have no information on the extent to which labour migration had formed part of the recent or historical work experience of Mabumba men and women, or on the effects this might have had on their economic behaviour.

In many rural communities, the effects of the long-term link forged by migration between town and countryside have been profound. This was the focus of Audrey Richards' work in the Northern Province in the 1930s (Richards 1939) and a salient feature of Moore and Vaughan's later study of the same area (Moore and Vaughan 1994). They found that some returned migrants had used their savings to invest in hybrid maize production, and were doing well, but other former town-dwellers found it difficult to settle to rural life. Men in particular were unsure what local economic activity was appropriate for them, and social problems arose when men could not find employment.

It is here that we begin to see the importance of a wider perspective on work or labour than that adopted in many time-allocation studies. Assessments of what work is may be shaped by experiences in different labour markets, especially where there are hierarchies between types of work, and if some work is subject to negative stereotyping. Some rural men may have had a long period in urban employment where jobs are more highly differentiated in terms of skill and pay. How do they adjust to dramatic shifts in labour markets implied by these rural-urban-rural movements? To what extent are decisions about rural work related not just to local labour markets, but to the biographical experience of work?

Conclusion

Policy-makers and practitioners working in rural sub-Saharan Africa are dealing with economies in which households struggle to survive through their own production, and by buying and selling in the market. Historically, there have been misunderstandings on many levels of the economic behaviour of men and women in these contexts. Agricultural development in the 1960s and 1970s systematically denigrated peasant farmers' knowledge of their own fragile environments, crop mixes, and husbandry practices. It was very hard for white educated agricultural scientists to accept that black illiterate smallholder farmers knew best.

This paper has focused on what I regard as another persistent area of deliberate, or unthinking, misunderstanding — namely, in judgements about what work is, and how it should be measured. Activists and researchers have lobbied hard to make policy-makers aware of the work that women do in rural sub-Saharan Africa — whether in farming, in other income-generation, or in essential household maintenance work — and to move away from Western stereotypes of man as producer and woman as homemaker. As a result, the concepts of economic activity have widened, and new concepts and technical tools have been developed to measure women's, as well as men's, work. Among these are time use surveys, which are a familiar tool used by gender and development practitioners in awareness-raising and planning.

This article has shown that time-use surveys may provide inadequate understandings of women's and men's work, in the absence of an understanding of the significance of the local context in which the work is done, including the relationship between farm and off-farm work, and of labour markets. If statistics on patterns of men's and women's activity are interpreted out of context, they can produce a distorted picture of the gender division of labour. Recent representations of women's and men's time-use in Zambia, which fail to examine how these differences between men's and women's activity arise, serve to construct an implicitly racist stereotype of African men as at best economically inactive, and at worst as idle. This stereotype arises from double standards. Mabumba men do little or no household maintenance work, in sharp contrast to their wives, but I have argued that this criterion for measuring men's workload is rarely applied in relation to public policy elsewhere. Mabumba men do appear to spend less time in work off the farm and in farming than their counterparts in other Zambian rural communities, but these differences are neither examined nor explained. Ideas of men as lazy have a long history in European discourses about rural sub-Saharan Africa — emerging wherever rural men resisted colonial labour regimes and coercive forms of rural development.

Although much has changed, the example of Allen's study, and its use in a recent World Bank policy document, suggests that work remains a value-laden concept. Researchers, policy-makers, and practitioners must indeed understand the different work burdens and productive contributions of rural men and women in sub-Saharan Africa, and elsewhere. To do this, it is necessary to be more sensitive to ways in which the researcher's own cultural concepts come into play, in the measurement and evaluation of work and in depictions of the gender division of labour.

Ann Whitehead is a social anthropologist who teaches social anthropology and gender and development at the School of Social Sciences, University of Sussex, Falmer, Brighton, BN1 9QN. E-mail: a.whitehead@susx.ac.uk.

Notes

1 The material in this article is taken from a longer paper that was presented in an ESRC seminar series on 'Men, Masculinities and Development' and at the 1999 DSA conference. My thanks to participants for their comments, and to Bridget Byrne and Caroline Sweetman for help in editing this version.
2 This phrase is taken from the title of Seigel's 1989 essay, on which the material in this section is based.
3 Fagan 1966 gives a general account of the Lamba in the nineteenth century.
4 For example see Cooper 1966; Thomas 1973, Palmer 1986, Overton 1986, and Johnson 1992.
5 This refers to the women's movement activists of the past three decades to distinguish them from the nineteenth and early twentieth century movements with similar aims.
6 The United Nations Statistics Division published *The World's Women: Trends and Statistics* for the UN Fourth World Conference on Women in Beijing in 1995. Updated statistics are available on the internet at www.un.org/depts/unsd. See section on 'statistics and indicators for the world's women'; section 5.6 contains data about time use.
7 Gatter 1993 contains an interesting account of the APRT approach, from the point of view of an anthropologist voluntarily attached as the social scientists to a team in Luapula.
8 There are a number of methodological problems with the ARPT studies, which there is insufficient space to discuss here. Data were collected on a two-day recall basis for the 12 hours between dawn and sunset. Where the total amount of activity recorded did not add up to 12 hours, the remaining hours were assigned to a residual category, termed resting.
9 See for example studies reviewed in Brown and Haddad.

References

Allen, J M S (1988) 'Dependent males: The unequal division of labour in Mabumba households', Luapula, ARPT.

Beijing Platform for Action

Blackden, M and C Bhanu (1999) 'Gender, Growth and Poverty Reduction: Special Programme for Assistance for Africa', 1998 Status Report on Poverty in sub-Saharan Africa, World Bank.

Brown, L R and L Haddad (ed.) *Time Allocation Patterns and Time Burdens: A Gendered Analysis of Seven Countries*, Washington, IFPRI.

Cooper, F (1996) *Decolonisation and African Society: The Labour Question in French and British Africa*, Cambridge University Press.

Evans, A and K Young (1988) 'Gender issues in household labour allocation: The transformation of a farming system in Northern Province, Zambia : A report to ODA's Economic and Social Research Committee for Overseas Research', London, Overseas Development Administration.

Fagan, B M (1966) *A Short History of Zambia*, London.

Ferguson, J (1990) 'Mobile Workers, Modernist Narratives: A Critique of the Historiography of Transition on the Zambian Copperbelt', *Journal of Southern African Studies*, 16: 385-412.

Gatter, P (1993) 'Anthropology in Farming Systems Research' in J Pottier, *Practising Development: Social Science Perspectives*, London, Routledge.

Harrison, E (1999) 'Men, Women and Work in Rural Zambia', paper presented to workshop on 'Working Lives, Men and Masculinities' in the ESRC Seminar Series Men, Masculinities and Gender and Development, University of East Anglia, September 1999.

Johnson, D (1992) 'Settler Farmers and Coerced African Labour in Southern Rhodesia, 1936-1946', *Journal of African History*, 33: 111-128.

Kumar, S. (1994). 'Adoption of hybrid maize in Zambia: Effects on gender roles, food consumption and nutrition', Washington, International Food Policy Research Institute.

Moore, H and M Vaughan (1994) *Cutting Down Trees: Gender, Nutrition and Agricultural Change in the Northern Province of Zambia 1980-1990*, London, James Currey.

Overton, J (1986) 'War and Economic Development: Settlers in Kenya, 1914-18', *Journal of African History*, 27: 79-103.

Palmer, R (1986) 'Working Conditions and Worker Responses on Nyasaland Tea Estates 1930-1953', *Journal of African History*, 27: 105-126.

Reardon, T (1997) 'Using evidence of household income diversification to inform study of the rural non-farm labour market in Africa', *World Development*, 25(5).

Richards, A (1939) *Land, Labour and Diet in Northern Rhodesia: An Economic Study of the Bemba Tribe*, Oxford University Press.

Seigel, B (1989) 'The "Wild" and "Lazy" Lamba: Ethnic stereotypes on the Central African Copperbelt', in L Vail, *The Creation of Tribalism in Southern Africa*, London.

Skjonsberg, E (1989) *Change in an African Village: Kefa Speaks*, West Hartford CT, Kumarian Press.

Stoler, A and Cooper, F (1997) Introductory essay in Cooper and Stoler, *Tensions of Empire: Colonial Culture in a Bourgeois World*, University of California Press.

Thomas, R (1973) 'Forced Labour in British West Africa: The Case of the Northern Territories of the Gold Coast 1906-1927', *Journal of African History*, 14: 79-103.

United Nations Statistics Division (1995) *The World's Women 1995: Trends and Statistics*, New York, UN.

World Bank (1992) 'Zambia — Agricultural Sector Strategy: Issues and Options', Agriculture Operations Division, Southern Africa Department.

Integrating gender needs into drinking-water projects in Nepal

Shibesh Chandra Regmi and Ben Fawcett

This article shows what project planners can do to ensure women's true participation in the design and maintenance of development projects, without increasing women's workloads, and with the aim of raising their status in the family and community, as well as challenging men's prejudice.

Women's empowerment is a prerequisite for development, as well as a question of justice. In almost all rural communities in developing countries, it is primarily the women, and sometimes girls, who collect water, protect water sources, maintain water systems, and store water. Women spend a significant amount of their time on these activities; they also determine how water is used, which has a direct impact upon their families' health. Women's pivotal role was recognised during the International Drinking Water Supply and Sanitation Decade, 1981-90, and has been widely discussed in the drinking-water sector since then. Numerous projects implemented in the past decade have made some provision to recognise women's roles in water collection and management, and to promote their participation in project activities, but such participation tends to be limited, and is often tokenistic. Water projects seldom focus explicitly on the need to promote an equal balance of power between women and men.

This article uses the framework of strategic and practical gender needs (Moser 1989) in the context of the drinking-water sector, to argue that understanding how these needs are linked is essential for making drinking-water projects sustainable. We argue that projects and programmes which aim to meet the practical needs of women, men, and children in communities must also focus on meeting women's strategic gender needs. The conceptual categories of practical and strategic gender needs (ibid.) refer respectively to immediate perceived necessities that women lack in a specific context, and necessities which would enable women to change their subordinate status in society: for example, to control their bodies, bear and rear children, own land and property, fight against domestic violence, claim equal wages, or change the sexual division of labour[1].

While Moser's concepts are well-known to development practitioners trained in gender analysis, they remain outside the conventional framework for planning and designing water projects. Drinking-water projects are nearly always carried out by engineers — most of them men — whose goal is the simple, and laudable, one of bringing adequate quantities of good quality drinking water closer to the homes of the target communities. Nevertheless, most development activists, including many with an engineering background, would agree that their work is ultimately concerned with the twin aims of enabling women and men to meet their practical needs, and enabling the marginalised groups within communities, including women, to fight against oppression and exploitation.

We consider a focus on women's strategic gender needs in development projects important for two reasons. First, the intervention may then contribute to greater gender equality in society; second, focusing on women's strategic gender needs is the only way to ensure that women's and men's practical needs are met fully and efficiently. In other words, only if women become active partners in the development process can societies be built where both women and men can thrive equally. The failure of many drinking-water projects to achieve a sustainable impact bears out the need to recognise that strategic and practical issues are linked in this way.

With this article, we aim to convince project planners of the importance of considering strategic gender needs and of developing ways of incorporating them into their plans, and to enable gender-sensitive impact assessment and evaluation work to pinpoint strategic gender needs which have gone unaddressed in projects where these issues have not been considered. The arguments presented in this article are based on the findings of a two-year research project carried out for a project in Nepal funded by the British Department for International Development (DFID)[2]. The research focused on the eastern, western, and mid-western regions, where both gravity-flow schemes (in the hills), and point sources (on the Tarai plains) were analysed. The field research has been supplemented by an extensive review of literature from similar studies world-wide. The main research techniques relied on participatory methods, including activity calendars, access and control profiles, social and resource mapping, transect walks and observation; semi-structured interviews using comprehensive checklists; and focus-group discussions.

Women's participation in planning and implementing projects

Much research (for example, INSTRAW and UNICEF 1988 and Fong et al. 1996) has shown that women from all groups within a community need to participate fully in project activities, to ensure that projects are effective in the long term. However, evidence from our recent Nepal research indicates that many drinking-water projects continue to bypass women in the planning, design, implementation, monitoring and evaluation process. Too often, projects and programmes are designed which pay little attention to the links between technical change and social relationships.

Women are seldom involved in essential planning activities, although, as primary collectors of water, they are likely to know much more than men about the seasonal availability of water from various sources, about the quality of water from those sources, and about individual and communal rights to use those sources, which can create conflicts after construction if they are not taken into account. In many cases, projects have proved ineffective in the long run, as women stopped using, or were unable to use, those sources. In all the communities involved in the Nepal research, women complained that their water-collection time significantly increased (nearly four or five times) after they received the improved water services. This is because the tap-stands and the tube-wells are located along the roadside, where they cannot bathe freely and wash their clothes used during menstruation comfortably, for fear of being seen by males. In order to avoid this, women in Hile village in east Nepal (which is in the hills and has a cold climate) carry water all the way to their homes several times each day, spending significant amounts of energy to do this. In three villages on the Tarai plain (Motipur, Magaragadhi, and Gajedi) women reported waiting until dark to undertake these activities. They said that they had not had this problem when they had used more distant traditional sources, where there was no chance of men being around. All these women also complained that the surveyors had not involved them in designing the tap-stands or tube-wells.

These findings echo other research in Nepal (Mustanoja 1998), in which women said they would like to increase the distance between the wall and the tap and to adjust the elevation of the platforms, to accommodate their *gagris* (watervessels which are carried on their waists), and stated that they do not find the platforms comfortable for washing laundry (Mustanoja 1998). Another research into the lack of women's involvement in water-project design found that hand-pump handles were either too long or too short, making them uncomfortable for women to operate, and sometimes causing injuries (IRC 1992).

If design and location of the new water systems are inappropriate, women are not likely to be interested in protecting them. For example, a poor woman from the lower ethnic group called Mallah in Gajedi remarked to us with frustration that she and many other women from her ethnic group who live in one location still spend nearly one hour collecting water from the new tube-wells. Yet women from relatively well-off families spend hardly five to ten minutes in this task, as their husbands were able to influence the installation of tube-wells to be close to their homes. She observed that this discrepancy makes her feel that women who are not benefiting from the improved water services should destroy the tube-wells, so that all women are then on equal footing in the community (personal communication, 1999).

In other cases, water systems cease to function because women have no control over them. Women's participation may be limited to some women being invited to meetings, to be nominal members of water committees, or, at the most, to take demanding and often tedious roles (van Wijk-Sijbesma 1985, IRC 1992). The few women on water users' committees may be selected by project officials in consultation with local men or the local NGO. As a result, these few women feel obliged to the male members of the committee and are reluctant to disagree with any decisions made by the men, regardless of whether or not those decisions are in women's favour. In Hile in east Nepal, the two women on the local water committee reported that they had not known for months that they had been selected. Because the male committee members had been instructed by the project officials to include two women in the committee, they had put the women's names forward as a token in order to activate the implementation of the water project. These women said that they have no chance to oppose what male members of the committee decide: they are not invited to participate in meetings, nor are they included in the sub-committee formed to monitor the project's progress (personal communication, 1999).

If women committee-members are selected through a democratic process — and particularly if they are elected by other women — the chances are high that they will be vocal, and concerned to guard women's interests, as they feel accountable to other women in the community. Similar principles apply in women's participation as pump and tap-stand caretakers: provided women are offered a proper role, and any training necessary to help them perform these tasks, they are highly likely to be most concerned about the proper use, and maintenance of water supplies because of their role as primary users. This is borne out by a study of the performance of women hand-pump caretakers in Bangladesh, which concluded that after 15 months of maintenance by women, the condition of the pumps was found to be as good as that of pumps maintained by trained project mechanics (Bilquis et al. 1991).

However, the male prejudice that women cannot contribute effectively to water projects because these are technical in nature, makes some involved in drinking-water projects think that it is more difficult to work with women. Various reasons may be given, such as their lower levels of literacy, lack of similar involvement in development projects, and social and cultural factors. In order to overcome these problems, the Nepal research has suggested the following points so as to increase women's participation in water supplies.

• Inclusion of both local men and women in the project activities. Clear explanation of both the short-term and the long-term benefits — tangible and intangible — of the project to both men and women, from the beginning so that all feel motivated, and so men are happy to see women participate.

• Gender training and awareness-raising for all. A key target group should be men who perpetuate negative stereotypes of women. Training should aim to show the benefits of women's participation in public life, challenging religious, traditional, and social attitudes which severely limit this. It should also motivate men to share women's work such as child-care and household chores, which is one major reason why women do not participate in development projects. Projects can also seek the help of local change agents to promote women's involvement, such as local leaders, respected elderly women, or school teachers.

• Promotion of women's employment in water projects, so women staff can work with women in the community.

• Allocation of adequate preparation time, including implementation of literacy and awareness-raising programmes either directly or through other agencies, to motivate women and build up their confidence, since preparing women to take up new roles should be an essential part of the ongoing development process.

• The use of participatory approaches, along with the presence of gender-sensitive men and women in the project team, to create an effective learning environment, even for illiterate people.

Changes in the gender division of labour

If drinking-water projects are to have a chance of being effective and of improving the lives of poor rural women, a focus on changing the traditional gender division of labour is essential. While improved water facilities are often assumed to lessen women's workload — and this is often a stated aim of projects — this may not be the case. As stated at the start of the article, the collection of water tends to fall entirely to women (unless they are sick, or, as in some South Asian communities, menstruating). In communities where the water source is very distant from the village, men may help their women so that they feel more secure, but this is likely to change when the new water source is activated. Water consumption also often increases once the water-source is nearer to home, requiring water to be collected many times a day. The time and energy expended by women may be almost the same as before.

Our research in Motipur, Magaragadhi, and Gajedi villages in western Nepal illustrate this point. Women work up to 18 hours a day here, while men usually work up to 13 hours. Apart from ploughing, which is considered to be a male reserve, there is hardly any regular activity which is performed exclusively by men; but there are many which are exclusively female. In their supposed rest hours, women knit, weave, and sew, while men spend their time drinking and playing cards. Men thought that their agricultural work, which is mainly ploughing and preparing the fields, is much harder and more difficult than that of women. In fact, women not only work longer hours but some of their activities, such as collecting fuel, fodder, and water, are at least as labour-intensive as men's work in the fields. In all communities, the women reported that they used to collect water four to five times a day, amounting to a total of 80 litres per family per day. But after they got water supplied near their homes, they fetched water 10-15 times, with households using nearly 200-300 litres of water a day.

Although women's active involvement in drinking-water projects is essential, it may be hindered by their workload. Women's triple roles in production, reproduction, and community management (Moser 1989) leave them with very limited time and energy to participate in project activities. Project planners and implementers may think that women are not interested due to their lack of participation, and proceed to design and implement projects without women's involvement, repeating the mistakes of the past. This is a vicious circle for both women and development.

Projects which aim to achieve sustainable improvements in water supply, as well as deeper social development, need to create an environment in which men are willing to share the work traditionally done by women. Without this, neither practical nor strategic needs will be met effectively. Two women members of the water users' committee in Gajedi village reported that they had attended only one out of ten local committee meetings held last year as the meeting place was far and there was no one to share their work at home. They said that although their husbands support them to participate in such meetings, they do not realise that this is impossible if the men do not share the work at home. These women suggested that the projects should focus more on how to motivate men to share women's work, rather than spending time in involving women in project activities which is never meaningful without men's sincere co-operation.

Changes in the gender division of labour not only enable women to participate in development projects and render them more likely to be sustainable; they also have a positive effect on women's health, which can lead to many other social and economic benefits for all family members. For instance, it can lead to improvements in family nutrition and health, because more time is spent on preparing food and looking after children (Jazairy et al. 1992); more time spent on income-generating activities may lead to increased income and other impacts (Van der Laan 1998, Curtis 1986), discussed further in the next section. Development activities where women's presence is essential for success benefit if they have more time to participate (Van der Laan 1998, Jazairy et al. 1992); in many societies where girls are prevented from going to school because they have to help their mothers, education may become possible (Curtis 1986, Aziz and Halvorson 1999); ultimately, as children witness the wider benefits of their parents sharing domestic work, a society with greater gender equality may emerge (Regmi 1999).

Some suggestions to help to bring about changes in the gender division of labour include the following.

• Water projects could initially emphasise that men should share the work of water haulage. Once this is achieved, gradual changes may occur in other activities.

• Awareness-raising through use of participatory methods of gender analysis, such as the preparation of activity calendars for both sexes. This process can help to overcome many patriarchal biases and should involve local authorities such as village heads, religious leaders, traditional healers, school teachers, and political leaders.

• Other gender-sensitisation activities, such as mass-meetings, film shows, workshops, and cross-cultural exchanges for local women and men, to highlight the negative effects of women's high workloads, and highlight the positive effects of both sexes sharing domestic work.

• The introduction of non-formal education programmes including literacy classes and technical skills training. (In Nepal, women's literacy rate is 25 per cent, compared to 55 per cent for men.) Many argue that these can empower women by offering them both practical skills and a chance to increase their confidence and self-esteem, a basis from which to challenge the apparent rigidity of social structures (*Waterlines* 1998).

• The introduction of income-generating activities targeted at both women and men. The realisation that women can share responsibilities for supporting the family can motivate men to share some of women's traditional work.

Increasing women's control over resources

Income-generation

Many water projects in developing countries aim to increase women's participation in income-generating activities or employment. The reasoning is that women will spend less time on collecting water, and will be able to invest this in producing for the market. Most focus on the practical benefits to the household of increased income, but there is also evidence that women's tangible economic contribution to the household is a key determinant of their status both in the household and the community (for example, Jazairy et al. 1992). Thus, women's involvement in income-generation may have a strategic impact on their status.

However, a key criticism to be made of the Nepal projects examined is that none had integrated income-generation into their activities. In the absence of such support, women who had saved time as a result of the water projects had taken on more knitting or weaving for the household, and more agricultural work, rather than income-generating activities. Women's increased agricultural work enabled men in the household to increase the time they already spent in migrant work as wage-labourers in urban areas. Male migration is seen as the key income-generating strategy for many in these areas, since households farm small land-holdings, and lack the agricultural inputs needed to produce for sale. Women in the project areas mentioned other barriers that prevent them from embarking on production for cash: it was not seen as profitable to produce processed foods such as jam, jelly, potato or apple chips, due to the lack of local markets, their inability to compete with producers in larger urban markets, and the fact that women's mobility is limited due to cultural factors. These findings are in line with other studies which show that a number of factors must be considered in planning any income-generating activity (Mayoux 1991); these include training offered to women, and the timing of such training, ensuring regular supplies of inputs and resources, easy access to markets, involvement of both men and women, provision for literacy and numeracy training, and so on.

Training offered to women in the Nepalese communities as part of water projects tended to focus on health, hygiene, and sanitation, and administrative skills including record-keeping; this has also been found in other contexts (Mustanoja 1998). These skills are not transferable, and the training women typically receive during such projects lasts only one week. In any case, women stand almost no chance of finding employment in a situation of high unemployment, and when the number of educated young people (mostly male) is increasing in every village. In contrast, technical training offered to men (such as masonry or latrine construction and maintenance) is always in high demand, both locally and outside the village. A rationale offered by water-project staff was that the job of a caretaker in a point source is easier than the job of a maintenance worker in a gravity-flow scheme (this is, however, inaccurate, since the tube-wells may need the same amount of time once they get older) (personal communication, 1999).

Community employment on water projects

Turning to the question of women's work within the water projects, the projects studied in Nepal tended to include women only in unpaid roles: water users' committee members, voluntary community health motivators and pump/tap-stand caretakers. However, a very small number of women who were relatively active in these capacities did report receiving an occasional income from them. In contrast, men were hired as maintenance workers. In the few instances where women were hired with men as wage-labourers — for example, during the construction stage of the Hile drinking-water project — they were paid lower wages than men. The women labourers said men should in fact have been paid less than them, as they spent time chatting and smoking cigarettes, while women are very sincere in their work (personal communication, 1999).

In all the communities, we observed that most women did not achieve any direct increase in control over household income from their participation in project activities. However, some women did report that they were consulted more by their husbands and other male family members when the men made decisions about household and capital expenditure. (This was true of all the women studied, irrespective of socio-cultural differences such as ethnicity, economic status, educational status, and remoteness of the area.)

The issue of payment is important when considering project sustainability: if there is no increase in income to compensate for the added responsibilities of being involved in managing water projects, women may decide not to take on this demanding work (Green and Baden 1994). It is critical to ensure that there are motivated, skilled caretakers and maintenance workers in the village. Women are especially important in these roles, as they are regular users of the service for domestic use, and the first to notice any defect in the system. Likewise, women are effective in the regular collection of water tariffs for operation and future maintenance. However, although water supplies with women caretakers may have a greater chance of sustainability, most of them work as volunteers, which places a satisfactory long-term operation in doubt; women may be unwilling or unable to give the increasing time needed as the equipment ages.

Women on project staff

Our research also focused on women who were formally employed by organisations involved in drinking-water projects. While the ratio of women to men is very poor at all levels — head office, regional office, and project office — the presence of women in senior positions and within the technical sector is negligible. The few women who were employed said that they cannot participate on an equal basis with men, and attributed their difficulties to bias about women in these 'hard' roles (personal communication, 1999). An example comes from the government district water-supply office in Dhankuta, eastern Nepal, where three women were recruited as water-supply and sanitation technicians. However, senior officers decided that women should not undertake labour-intensive activities in the field, and so these employees have been re-assigned to perform administrative tasks (personal observation, 1999). It is essential that water projects make every effort to involve women in paid positions, pay them at equal rates to those paid to men, and provide them with training in areas which can give them income in the future, which are necessary not only for the sustainability of practical benefits but also for greater gender equality and justice.

The links between practical and strategic needs are illustrated in the Nepal research in matters such as income-earning, status in the household, and project sustainability. Women members of water committees in Magaragadhi and Gajedi agreed that as long as they were not earning income from the project, their husbands would not appreciate their opinions. One participant stated:

'Since we have been asked only to do the non-technical activities, and [have] not been provided [with] any technical training, we are not in a position to make any income even after the completion of the projects, unlike the men, who have received training on latrine building construction and masonry. This is why we have been demotivated to hold any meetings for the last couple of months and to take any initiatives yet to resolve the problem of malfunctioning tube-wells which is increasing over the years' (personal communication, 1999).

Assurance of benefits to women in marginal groups

For women to benefit fully from water projects, they need to be seen as individuals whose gender identity links to age, ethnicity, and economic class. In the Nepal research, a widow in Magaragadhi village, whose husband had died years ago, reported that she was never invited to participate in any meetings including that which dealt with the siting of tube-wells. Since there was no one to speak for her, and she could not voice her concern due to lower socio-economic status, the tube-well in that community was installed far from her home. This made it difficult for her to go to do the agricultural wage labour which is the chief means of survival for herself and her three children. She asked,

'will there ever be a time when such poor women like me, who cannot voice their concerns, are also equally treated in the community?' (personal communication, 1999).

Water tariffs raise particular issues for women in different household forms. Tariffs are often set at equal levels for all the user households, without considering factors like the number of users in the family, the number of income earners, and gender relations within the household. In many male-headed households in the Nepal research, men said that paying the water tariff was women's responsibility, as women deal with water. However, some women reported difficulty in paying the fee, since they do not have control over income (Regmi 1999). The problem of paying water tariffs may be particularly acute among female-headed households, where women have control over the income, but more limited resources.

For example, in one meeting about the collection of water tariff in Gajedi village we found that the tariffs were mostly paid by women; women from female-headed households were among the defaulters. A decision was taken in the meeting that if the defaulters did not pay their dues within 15 days, they would not be allowed to use the tube-wells. Two of the defaulters participated in the research; both women were very poor, living a hand-to-mouth existence; one had two children and the other three. Their small pieces of land were insufficient for their survival, and they were labouring on others' farms or as servants for wages in kind. They did not know what they were going to do if they were banned from using the tube-well. There is a danger that families in such situations may return to unhygienic water resources, risking their health (Evans 1992, Fong 1996).

Focusing on improving women's status

It is important to have an explicit focus on improving women's status. Unless the root causes of women's subordination are identified and addressed, and their genuine needs are prioritised, development projects and programmes — including those in the water sector — which involve women with the aim to empower them will not lead to significant and lasting improvements in their lives. This article has raised a number of issues to consider in this regard. Even if projects have a direct impact in improving the lives of only a few women, they can have long-term multiplier effects on other women in the family and community, acting as role models.

Conclusion

The lack of adequate water supplies is a major problem in developing countries. The problem is worst in rural areas and among poor households in urban areas, as these populations cannot afford the cost of either installation or operation and maintenance. This article has discussed the great need to involve women in the management of water projects, so that such projects become effective in reducing people's hardship. In addition, all development activities, including water-supply improvements, should be concerned with improving the lives of women in strategic as well as practical ways — in other words, changing the status of women and increasing their confidence. These aims are not mutually exclusive; rather, they reinforce each other, since a country's development depends on the active participation of both women and men in the overall development process.

Meeting women's strategic gender needs does not demand a huge financial investment from those designing water projects. It does, however, require a genuine commitment from the people involved at all levels, and budgetary provision to build the capacity of all involved in the sector, and raise their awareness. In turn, the fulfilment of women's strategic gender interests can contribute to the sustainability of water projects, ensuring that governments and international funding organisations do not waste their huge capital investments, and achieve sustainable human development.

Shibesh Chandra Regmi is former Executive Director of New ERA, a research firm in Nepal established in 1971. He has worked in the development sector for 17 years, focusing on health, rural development, women and gender, and participatory planning and is currently undertaking PhD research at the Institute of Irrigation and Development Studies, University of Southampton, UK. Email: scr1@soton.ac.uk and info@newera.wlink.com.np.
Ben Fawcett is the Co-ordinator of the Engineering for Development Programme at the Institute of Irrigation and Development Studies at the University of Southampton, UK. He has over 17 years' experience in engineering and international development, especially in environmental health and programme management in Asia and Africa.
Email: bnf@soton.ac.uk

Notes

1 For a discussion of the distinction between the two and the implications of this, see March et al. 1999.
2 The research included Motipur and Magaragadhi drinking-water projects implemented by Nepal Water for Health (NEWAH), a leading Nepali NGO which is mainly funded by WaterAid UK; Gajedi drinking-water project imple-

mented by Rural Water Supply and Sanitation Project (RWSSP) funded by FINNIDA; and Hile drinking-water project of the Fourth Rural Water Supply and Sanitation Sector Project (FRWSSSP), funded by the ADB/Manila and implemented by the Department of Water Supply and Sewerage, the lead government agency in the water-supply sector in Nepal.

References

Aziz, N and Halvorson, S (1999) 'Women's involvement: A switch in thinking, Hoto, Pakistan', in *Community Water Management*, Participatory Learning and Action (PLA) Notes, Interntional Institute for Environment and Development (IIED), London.

Bilqis, A H et al. (1991) 'Maintaining village water pumps by women volunteers in Bangladesh', in *Health Policy and Planning*, Vol. 6, No. 2.

Cleaver, F and Elson, D (1995) *Women and Water Resources: Continued Marginalisation and New Policies*, International Institute of Environment and Development (IIED), Gatekeeper Series No. 49.

Cleves Mosse, J (1993) *Half the World Half a Chance: An Introduction to Gender and Development*, Oxfam, Oxford.

Curtis, V (1986) *Women and the Transport of Water*, Intermediate Technology Publications.

Elson, D (1991) 'Structural adjustment: Its effect on Women', in C March and T Wallace (eds.) (1991).

Evans, P (1992) 'Paying the piper: An overview of community financing of water and sanitation', IRC International Water and Sanitation Centre, Occasional Paper No. 18, IRC, The Hague.

Fong, M S; Wakeman, W; and Bhusan, A (1996) 'Toolkit on gender in water and sanitation', Gender Toolkit Series No. 2, The World Bank, Washington, DC.

Green, C and Baden, S (1994) *Gender Issues in Water and Sanitation Projects in Senegal*, Institute of Development Studies, Sussex.

Gwen, I C (1996) 'Gender perspectives of the water ssector', in *Water Sector News*, No. 4, pp. 5-6.

INSTRAW and UNICEF (1988) 'Women and water supply and sanitation', national training seminar held at Kadugli, Sudan, 16–21 January 1988, UN International Research and Training Institute for the Advancement of Women, Santo Domingo.

International Reference Centre (IRC) for Community Water Supply and Sanitation (1992) *Women, Water, Sanitation: Annual Abstract Journal*, No. 2, IRC, The Hague.

Jazairy, A; Alamgir, M; and Panuccio, T (1992) *The State of World Rural Poverty: An enquiry into its causes and consequences*, International Fund for Agricultural Development (IFAD), Rome.

Joshi, Deepa (forthcoming 1999) 'Gender issues in the management of water projects in UP Hills of India', University of Southampton.

Longwe, S H (1991) 'Gender awareness: The missing element in the Third World development project', in C March and T Wallace (eds.) (1991).

March, C et al. (1999) *A Guide to Gender-Analysis Frameworks*, Oxfam, Oxford.

March, C and Wallace, T (1991) Changing Perceptions: Writings on Gender and Development, Oxfam GB, Oxford.

Mayoux, L C (1991) 'The poverty of income generation: A critique of women's handicraft Schemes in India', in C March and T Wallace (eds.) (1991).

Moser, C (1993) *Gender Planning and Development: Theory Practice and Training*, Routledge, London and New York.

Mustanoja, U M (1998) 'Gender Analysis and Integrated Gender Plan', Rural Water Supply and Sanitation Project (RWSSP), Plancenter Ltd., Ministry for Foreign Affairs, Department for International Development Cooperation, Lumbini Zone, Nepal.

Penny, A (1991) 'The forward looking strategies', in C March and T Wallace (eds.) (1991).

Philippa, Hill (1998) 'Femconsult newsletter: Men, women and water participatory approaches', in *Water Management*, No. 1, pp. 2-14.

Population Reports (1998) Series M, No. 14, Special Topics.

Pugansoa, B and Amuah, D (1991) 'Resources for women: A case study of the Oxfam sheanut loan scheme in Ghana', in C March and T Wallace (eds.) (1991).

Regmi, Shibesh Chandra (forthcoming 1999) 'Gender issues in the management of drinking water projects in Nepal', University of Southampton.

Thresiamma, M (1998) 'New skills, new lives: Kerala's women masons', in *Waterlines*, Vol. 17, No. 1, pp. 22-24.

van der Laan, A (1998) 'Case study: A participatory water supply scheme on a tea estate in central Sri Lanka', The Hague, The Netherlands.

van Wijk-Sijbesma, C (1985) 'Participation of women in water supply and sanitation: Roles and realities', IRC for Community Water Supply and Sanitation Technical Paper, No. 22, IRC, The Hague.

Wadehra, R (1991) 'Breaking the mould: Women masons in India', in C March and T Wallace (eds.) (1991).

Wakeman, W; Davis, S; van Wijk, C; and Naithani, A (1996) *Sourcebook for Gender Issues at the Policy Level in the Water and Sanitation Sector*, International Bank for Reconstruction and Development/ The World Bank, Washington, DC.

Wallace, T (1991) 'Case Studies of Ways of Working with Gender', in C March and T Wallace (eds.) (1991).

Waterlines (1998) Vol. 17, No. 1.

Structural adjustment, women, and agriculture in Cameroon

Charles Fonchingong

This article appraises the impact of economic structural adjustment programmes (SAPs) on the agricultural activities of women's groups in Cameroon, and explores women's ways of coping with the decline in individual and family income and the loss of public services.

In Cameroon, agriculture is part of a livelihood strategy to safeguard a family's food security, health, and children's education. Since the introduction of SAPs in the late 1980s, some women are spending more time in agriculture to offset declining incomes and pay for a range of social services, growing crops for sale, barter, or subsistence, while others combine farming with entrepreneurial activities.

Structural adjustment in Cameroon

The Cameroonian economy recorded a high growth rate between 1975 and 1983. This successful economic performance — the result of a rise in investment, exports, and consumption — followed a period of intense development efforts, during which nearly all economic indicators were favourable. But from 1987, the economy contracted considerably, and progress towards improving the welfare of the population and meeting their basic needs more effectively was compromised.

SAPs were introduced to many African and Asian countries during the 1980s by the World Bank and IMF; the aims were to salvage the deteriorating economies of these countries, through redefining the role of the state, reforming the civil service, and rehabilitating public enterprises and parastatals, in order to foster efficiency and stimulate growth. The process of structural adjustment was begun in 1987.[1] In 1994, the CFA franc (Cameroon's currency) was devalued; thereafter, public sector salaries were slashed, and a massive retrenchment of public service workers ensued. Currently, there are 180,000 public service workers, about 14 per cent of the total employed (Ministry of Economy and Finance 1999). UNICEF (1993) notes that women are the first to lose their jobs in periods of retrenchment, and that they become family breadwinners when their husbands are retrenched.

At the time of writing (August 1999) Cameroon is preparing to sign its third agreement with the IMF (a Stand-by Agreement of Loan Disbursement) which contains some modifications to the package of economic adjustment measures undertaken so far, and emphasises the stringent management of public expenditure. To date, Cameroon has carried out the following measures as part of economic adjustment: controlling government expenditure through cutbacks on public spending; restructuring budgetary expenditure on technical services, especially agricultural research and extension; promotion of cash crop production; rationalising the selection of public investment projects; restructuring and increasing revenue through fiscal reforms; and settling the government's domestic arrears[2] (Ntangsi 1998). Government statistics put Cameroon's growth rate at 4 per cent (Ministry of Economy and Finance 1999), but this still remains to translate into standards of living.

The research

This article explores some of the various problems facing rural and urban women and their families under economic adjustment, and the role of women's groups in helping women to cope with the economic crisis which adjustment has caused. While rural households can try to stave off hunger and malnutrition through subsistence cultivation; in contrast, in urban areas, small parcels of land that used to be available for cultivation are being eaten up by urban expansion. The impact of the crisis and people's capacity to cope vary from one social group to another. However, overall, the study indicates that the groups most vulnerable during economic adjustment have been the urban poor, women, old people, children, and those living off their savings or on fixed incomes. This article will focus on women, and compare their experiences in rural and urban areas.

The article draws on research conducted between April and June 1999 into the role of 25 women's groups in both rural and urban areas, in Cameroon's north-west and south-west provinces. The research explored the aims and objectives of the groups, the impact of group activities on members, and any problems they had experienced as a result of structural adjustment. The groups involved in the research were very different in structure: some had formal set-ups, with a president, secretary, and treasurer; these ranged in size between 10 and 20 members. Other groups without formal structures were particularly common in rural areas; these had between 30 and 40 members.

The primary method of research was a survey, but it was augmented by in-depth semi-structured interviews and focus-group discussions with group members, to discuss the benefits of belonging to a group. However, much vital information and data were gathered through conversations which gave respondents from very different backgrounds the opportunity to express themselves freely, and to discuss their experience of fighting the crisis.

Rural agriculture: Changes and challenges

For rural dwellers in Cameroon, agriculture is the backbone of livelihoods. Men used to engage in cash crop production, and women were chiefly concerned with food-crop production, but women reported that the economic crisis has changed the way they work with men. In Cameroon, as in other countries, structural adjustment measures aim to encourage the production of cash crops for export, to generate more foreign exchange and render the country better able to service its external debt payment. Crops grown for sale include bananas, palm oil, coffee, cocoa, and groundnuts. (Coffee is mostly cultivated in the north-west region of Cameroon, while cocoa is cultivated in some parts of the south-west province).

In comparison to the way of life before economic adjustment, women felt that there is now hardly any dividing line between men's and women's work in farming. There clear division of labour, with men concentrating on cash crops and women on food crops, which existed before the crisis has now changed. The distinction between cash crops and food crops has become blurred in some cases; for example, some women and their husbands are involved in the cultivation of food crops which are exported to neighbouring countries like Gabon, Equatorial Guinea, and the Central African Republic. Cash crops may be intercropped with subsistence crops on the same land. One response to economic pressures has been that everybody now works on the land to ensure survival.

However, there are tasks which are still commonly done by men, and food-crop production remains women's concern. Some respondents said that men tend to clear the fields and prepare the land for planting, and also help in planting, harvesting, and applying fertiliser. They also guard crops in areas where they are not safe. Women stated that they are still seen as responsible for feeding the family, and therefore have primary responsibility for subsistence crops.

Both cash and food crops may be grown on the family farm, and also on any other parcels of land which can be found for cultivation. Sometimes, cash crops may be intercropped with crops intended for subsistence. Commonly, though, cash crops are being grown on all available pieces of family land, and subsistence agriculture is pushed to small, and often remote, plots of marginal land. While the returns from these plots are small, it is time-consuming to cover the distance separating one such plot from the others. Most of the women interviewed were taking about four to six hours to trek from one farming plot to another. Women reported visiting each plot about three times a week during peak periods, and once a week in the slack season.

Some women who participated in the research thought that their continuing dependency on men for access to land hampered their agricultural efforts. Cameroon's pattern of land tenure is culturally determined: rights to use land are assigned by the tribal chief or village authority, and the male family member makes the decisions on land-use. To lack clear title to land is to be dependent on those who control it, even if women's rights to use the land are recognised (Young 1993). Half a century ago, it was observed that in Cameroon, 'men own the land, women own the crops' (Goheen 1996). Most women involved in farming produce food both to provide for their families and for sale: rural and urban food supplies are dependent on the food they grow. In addition, land can serve as collateral to enable women to get credit and develop their agricultural and other activities. Constraints on women's food production are therefore likely to have a negative impact on the health and wellbeing of a great part of the population, a risk which should not be ignored (Visvanathan et al. 1997).

The women I interviewed felt that their workloads had increased tremendously in recent years. In rural areas, most women farm with basic tools, and have no access to agricultural inputs. In addition to cultivating crops, women tend animals. Many are now involved in income-generating activities outside the home, including selling food crops in local markets or, if they are close enough to urban areas or if transport is available, in urban markets. However, because urban dwellers' incomes have declined, they can only get low prices for their foodstuffs in urban markets.

Some women reported that their contribution to the household is now more visible. While men were formerly seen as family breadwinners, both sexes now share this role. This is reflected in greater control of income: more than 60 per cent of the women interviewed said that they now manage household income, usually because they are better managers and carers for the family. Research in other contexts argues that women are motivated principally by the needs of their children their households, whereas men are motivated to invest time and money outside the household, in male-dominated networks and business partnerships (Rowlands 1995). However, women in my research complained that men squander the resources at their disposal on alcohol consumption, women, and social activities.

Agriculture and urban livelihoods

In urban areas, drops in household income and rising costs of living as a result of economic adjustment have forced women and men to eke out a living from agriculture. Food-stuffs eat into people's earnings, especially since the CFA devaluation of 1994. A higher proportion of household money is required to pay for medical care and education. Schools have high drop-out and low enrolment rates due to hardship resulting from economic adjustment. In the course of the research, I found that many girls were taking part in informal sector activities to supplement household income, rather than attending school.

Intercropping and share-cropping are common practices in urban agriculture due to scarcity of land. Organic waste is often used instead of fertiliser, but urban women who are able to afford agricultural inputs have relatively easy access to them. They may also be able to obtain credit from government and non-government sources, and from *Njangi* (rotating savings and credit) groups. Most of the women in urban areas said that the land they used for farming was rented, and their grip on it was temporary. Plots of land which have not yet been developed are offered for this purpose. An example is Nkwen, a rural area near the town of Bamenda. The Nkwen women's group farms vegetables on marginal lands rented from landowners who have unexplored land on the fringes of town. The women's group did not consider it sensible to invest heavily, since the group was not sure of retaining the piece of land during the next farming season. Landlords were compensated in cash or in kind.

In urban areas, members of women's groups listed many income-generating activities taken up in response to economic adjustment. Both sexes are commonly involved in informal entrepreneurial activities, but which activities are undertaken by women and men depends on class, occupation, age, education, and at times tribal affiliations, and women's trading activities tend to be on a smaller scale than those of men.

Some women stated that they are employed (as clerical workers, tailors, cleaners, hair-dressers, bar managers), but many also need to work in the informal sector to make ends meet. Petty trading of food and consumer goods is a key strategy for women: they buy foodstuffs (including cocoyams, plantains, beans, and vegetables) or stationery and other small goods from wholesalers ('buyam-sellams') and resell these, or cook and sell food at vantage points in town. Members of women's groups reported that most men (whether employed or not) are also involved in petty trading, street hawking, or selling a variety of goods as itinerant traders.

Coping with adjustment

Balancing the workload

Since women's lives straddle the reproductive and productive spheres, they absorb much of the pressure of structural adjustment. When women return from cultivating crops for sale, they continue farming on the homestead (Young 1993). Because they spend more time producing crops for sale, in informal sector activities, and providing family health-care, women are left with less time to carry out subsistence agriculture. Despite this, women reported putting in more hours on food-crop cultivation on their small land holdings, in combination with running the household, and collecting water and fuel, which adds to their laborious agricultural tasks.

In both urban and rural areas, women have been putting in more hours on food-crop cultivation since economic adjustment to compensate for the diminishing amount of available land and the lack of inputs. For urban women, who are also undertaking entrepreneurial activities, this has been a major drain on their time and energy. During 1998-99, land productivity in urban areas has gone from an average of about 40 per cent to 70 per cent, and in rural areas land productivity has risen from about 70 per cent to 85 per cent. In urban areas, the impact of reform measures can be seen by the increased number of women engaged in food-crop production to satisfy the practical needs of the household.

Some women said that the time they spend with their children is much reduced, with the consequence of reduced standards of care. Women who are breast-feeding children are faced with the most acute problems; in extreme cases, children have been abandoned. In most female-headed and some male-headed households, daughters take over the management of the family while the mother farms. Most members of women's groups said that their schedules were overloaded as they shuttle from the farms to the market and to meetings and other group activities, especially during the weekend. Women explained that they cope with the growing workload by foregoing recreation, reducing their hours of sleep and leisure time, and having fewer social outings. Since Sunday is a day set aside for rest, it is a good day for group meetings after church services in the mornings.

Counting the gains of group membership

All the women's groups felt that getting together and sharing resources was a good way of sustaining a livelihood in times of economic crisis, and of ensuring the survival of their families, although there were wide differences between the kinds of work they had embarked upon. Most women felt that their standard of living had fallen due to the crisis, but was now improving. They felt that this was due to the exchange of information, knowledge, and other resources in the groups, and second, due to the fact that they had engaged in more agricultural activities than before the crisis. They saw this as a critical factor which had helped them meet the basic needs of their families.

Most women considered membership of their groups as a source of strength, helping them to stem the decline in living standards and in purchasing power. Some activities of rural and urban groups have been highlighted above: they are involved in collective farming and micro-processing and selling of food stuffs. For example, the Babungo women's group, the Batibo women's group and the Manyu women's group of north-west and south-west provinces process cassava and sweet potato into flour, soya beans into soya milk and other products. The Nkwen women's group, north-west province, is involved in shared micro-enterprises, including raising small ruminants and pigs, keeping poultry, and making textiles. Some of the groups make soap and detergents, and urban groups also focus on credit provision. Some groups buy necessities such as palm oil collectively and resell to their members at moderate prices.

In group meetings in both rural and urban areas, women commonly take part in savings schemes. During the week, women struggle to save money from their sales of produce for the meetings on Sunday. The schemes function on a 'thrift and loan' basis, where members save in rotation; savings are distributed via a ballot at the meeting, or according to the gravity of the problem faced by a member. Loans have a minimal interest charge, and are repaid after a specific period. In some groups, members get their savings at Christmas, when they need to buy extra commodities.

Meetings are also a forum for exchanging ideas about women's agricultural work, and discussing problems related to their subsidiary activities like marketing and processing foodstuffs. Because women can no longer afford to buy imported grains and other food items due to their high prices, they have to convert what they produce for their own consumption. For example, soya bean is produced on a large scale and processed into milk, flour, and other products, for sale and for home consumption. Urban groups have an edge over rural groups in processing, since they are more likely to have access to improved technologies, and urban selling can occur on a daily basis, whereas in rural areas markets are normally held once a week.

The Kongadzem[3] women's group

The Kongadzem group was set up by a few women in 1994 during the worst of the economic crisis, with the aim of improving the living conditions of rural women in the Bui division of Cameroon's north-west province. The first members recruited others, and the group focused on agriculture, although a few members were more interested in small business. By pooling their resources, women in the group have gained access to credit from several sources[4], which they have used to purchase agricultural labour-saving devices, such as corn mills, mechanical pressers for processing garri[5], cassava graters, oil pressers to extract oil from palm kernel, drying ovens, and wheel-barrows for transporting produce in remote areas.

Members of Kongadzem said that they had increased their food crop production as a result of collective farming and training visits organised through the group, which equipped them with skills in farming and livestock-rearing techniques. Like other women's groups studied, the members benefit from loans through the *Njangi* scheme. They rear livestock with the aim of fighting malnutrition in the community by augmenting protein intake. The first-born female offspring of an animal allocated to a group member is repossessed by the group, and given to another member.

Economic hardships are alleviated in several ways: for example, members receive training in book-keeping skills. Members contribute money to help others in financial need during occasions like 'cry-die' (rituals on the death of a close relative) and 'born-house' (rituals on the birth of a baby). These ritual activities are meant to cushion community members from pain and hardship, and to encourage members to forge ahead amid difficulties. Also worth noting is the policy of the 'trouble bank', which was a feature not only of Kongadzem, but of most other groups which participated in the research. The 'trouble bank' is meant to rescue members when they may have difficulty in meeting expenses, for example, in times of sickness. Women reported that they gain emotional strength from membership of the group, which inspires them to renew their efforts in order to sustain their families.

Conclusion

This article has attempted to shed light on the coping strategies adopted by women as a result of structural adjustment. Women's workload has increased since they are now farming for both cash and subsistence, using fragmented plots which are often distant from home, and yield poor returns. Land for food-crop cultivation has become increasingly scarce, and inputs have become increasingly unaffordable. The money which comes into the household from the sale of crops is insufficient to compensate for higher costs of living and social services. Women asserted that what money is available needs to be within their control, to ensure that it reaches the family.

Women in the third world now carry a double, even triple, burden of work as they cope with housework, child-care and subsistence food production, in addition to an expanding involvement in paid employment (Momsen 1991). In addition, women everywhere say that they work longer hours than men, meeting responsibilities at home in addition to productive work outside.

How women cope with economic crisis is crucial to the success of development policies in the third world. In the 25 women's groups in this research, women have devised strategies to deal with the crisis, but they need the support of organisations which can offer them incentives in the form of affordable agricultural inputs, credit, and other vital resources that will create an enabling environment for them to operate in. Women in rural areas seem to be coping better than urban dwellers, since they have more food crops to sell. Pressure on land in peri-urban areas means that meeting subsistence needs is very difficult; employment is scarce, and the informal sector is overcrowded. Urban women involved in the study spoke of extreme cases where women and girls resort to prostitution; they also attributed a high degree of delinquency among boys and men to an inability to cope with life in this setting.

Charles Fonchingong lectures in the Department of Women and Gender Studies at the University of Buea, Cameroon, Fax: +237 43 25 08, E-mail: ubuea@uycdc.uninet.cm

Notes

1 To deal with its deteriorating finances, the government first launched an economic stabilisation programme in 1987 to restore a budgetary balance; a first agreement was signed with the IMF in 1988. A second agreement was approved in 1997.
2 Debts owed to Cameroonian citizens, especially suppliers and contractors.
3 'Love all' (in the Banso language of Bui, north-west province of Cameroon).
4 The NSO Women's Cooperative Society; the Investment Fund for Communal and Agricultural Micro-projects (FIMAC), a loan scheme sponsored by the World Bank; the Association for Women's Information and Coordination Offices (AWICO), an NGO that facilitates loans to women's groups in the north-west province (known as the Women's Information and Coordination Forum, WICOF, in the south-west province).
5 Foodstuff made from cassava.

References

Commonwealth Group of Experts (1991) 'Women and Structural Adjustment', Commonwealth Secretariat, London.

Friedrich Ebert Stiftung (1997) *La Privatisation des Monopoles de Service Public Au Cameroun: Evolution et Enjeux*, Institut Supérieur de Management Public, Editions Saagraph, Cameroon.

Goheen, M (1996) *Men own the Fields, Women own the Crops: Gender and power in the Cameroon grassfields*, University of Wisconsin Press, USA.

Manga, E (1998) *The African Economic Dilemma: The Case of Cameroon*, University Press of America, USA.

Momsen, J (1991) *Women and Development in the Third World*, Routledge, UK and USA.

Ntangsi, M (1998) 'The structure of the economy of Cameroon', lecture notes, Department of Economics and Management, University of Buea, Cameroon.

Rowlands, M (1995) 'Looking at financial landscapes: A contextual analysis of ROSCAs in Cameroon', in Ardener S, Burman S (eds.) *Money-Go-Rounds: The Importance of Rotating Savings and Credit Associations for Women*, Berg Publishers, Oxford.

UNICEF (1993) *Women's and Girls' Advancement*, United Nations, New York.

Visvanathan, N et al. (eds.) (1997) *The Women, Gender and Development Reader*, Zed Books, London.

World Bank (1979) *Recognising the 'Invisible' women in Development: The World Bank's experience*, World Bank, Washington DC.

Young, K (1993) *Planning Development with Women: Making a World of Difference*, Macmillan, London.

Interview
Are genetically modified crops a new development?

Koos Neefjes and Penny Fowler

Are genetically modified (GM) crops a new development?
KN: Yes, in the sense that they are the result of new ways of genes, which have been developed over the past two decades. These technologies permit manipulating plants at a pace that nature can't achieve, for example bringing genetic materials of fish into crop varieties. They can transfer a gene across species, or from the animal kingdom to the plant kingdom. (It does happen in nature — for example, oil came from microbes.) These technologies are mostly being developed and controlled by private companies, whereas in the green revolution of the 1960s and 1970s, most technology was in the hands of public organisations. The products of gene-technologies, and also the technologies themselves, are rapidly being patented in the United States, and increasingly in other countries, including the European Union. Patents are a far more powerful means of defining rights to these technologies and their products than, for example, breeders' rights, which only recognise rights over the products of plant breeding, rather than the technologies or individual genes.

What is the argument for GM crops as a solution to global food insecurity?
PF: Some biotechnology companies argue that GM crops are the answer to a pending crisis in world food supply. We do indeed face a growing world population, and there are different views about levels of agricultural productivity and how they will change over time, but most commentators consider that global food production will keep up with global population growth.

I think there are two important points. The first is about total food availability at a global level: there is already enough food available to feed everybody if the distribution of assets was more equal. Food security is more a question of access to productive assets and income-earning opportunities, so that poor people can buy or grow their own food, than of global food supplies. The second point is whether GM crops offer opportunities to address food security concerns in different regions of the world. While there are certainly some possibilities for producing GM crops or products that might be of benefit to developing countries' local food security, such as salt resistant or pest resistant crops, there is very little financial incentive for private companies to invest in these areas. Donor governments and agencies should commit resources to investment in research into these opportunities.

KN: Another argument used in favour of GM crops for global food security is that environmental degradation is resulting in lower yields. Furthermore, labour productivity in farming is declining too; in the absence of capital investments and alternative employment this will lead to increased rural poverty, possibly food poverty. A further argument is on grounds of scientific progress — technologists argue that this is simply a more powerful technology than anything we have seen before. They have no understanding that food security depends on more than the technologies of actually producing it. That's another debate in itself.

What impact will GM agricultural crops have on the food security of disadvantaged groups in different regions?
PF: I think that's an unknown quantity at the moment, because it depends on how the technology is developed, who controls that development, and what their objectives are. A very small number of companies control the technology and since their activities are driven by profit, this makes the possibilities of generating positive uses of the technology in the public interest seem less likely. A lot depends on finding incentives to promote the development of those GM crops and products which are of interest to small farmers, and low-income consumers, by either the private or the public sector. There are some potential benefits to consumers. For example, if use of GM seed does reduce production costs, there is a chance that you might see lower prices for some basic food items for consumers in the future.

KN: If you only look at the negative effects of the first green revolution (because there *were* positive effects, in terms of increased production) they were increases in inequality, increased dependency. It is quite logical to expect similar effects with GM technologies, but they are going to be worse because these are much more powerful technologies, and the issue of gene and crop patenting and corporate control of both the technology and its products (i.e. seeds), combined with the fact that the same companies sell designated pesticides and herbicides, makes the potential for farmers' dependency enormous. However, there is no evidence about what will happen yet, because these patents have not been applied on any real scale, and the GM seeds that are being patented have not been commercialised beyond a few countries, notably the USA, Canada and Argentina. For the moment, the debates in Europe, India, Brazil, and elsewhere are mainly about allowing field testing of GM crop varieties, and imports of GM foodstuffs, whereas the commercial release is envisaged to take off in a few years from now. So it is all guess-work at present.

PF: There is a need not only to consider ownership, investment, and the direction in which GM technologies are developing, but also to ensure that proper regulatory frameworks are in place. Whether you are talking about the technologies being applied in the countries where they are developed, or about exporting them to other countries, mechanisms need to be in place to ensure that thorough and effective risk assessments are carried out in relation to public health, the environment, and also socio-economic impacts. This isn't the case in many Northern countries, let alone in Southern countries. It's fundamental that a precautionary approach is adopted to the development and application of GM technology.

Do you think consumer concerns that the crops may be a threat to health are valid?
KN: Yes. The probability that something may be wrong is tiny, but if it is, the consequences will be serious. We may see new and increased instances of allergies among consumers; plant viruses could transfer to gut bacteria, new human viruses could develop through recombining DNA, and many scientists fear increased resistance against antibiotics. The environmental risks could have enormous effects on human life, apart from those on wildlife, and on farming potential.

PF: Given the uncertainty over some of the health risks at the moment, the key issue is that consumers should definitely have a choice over whether or not to eat GM foods, for example through GM food labelling schemes. That has implications for policy decisions regarding international trade regulations, and the extent to which these regulations threaten the ability of governments and consumers, in North and South, to make informed decisions about importing and eating GM foods.

KN: If regulations on labelling allow European consumers to choose, maybe North American farmers will not manage to sell as much GM food, so it might well go to developing countries, possibly even through the food-aid systems. So we might end up with the double standards of European consumers refusing the food, and poor people in Africa having to eat it. This is an ethical and moral question.

What impact do you think patenting of new technologies has on people in developing countries, and particularly on Southern farmers?

PF: There are three key concerns. The first is whether it is appropriate for private companies to have a monopoly in terms of intellectual property rights over the genetics of traditional crop varieties, in order to use them to produce GM crops or products in future, without recognising the contribution farmers have made over centuries in developing those traditional crop varieties. The problem is that patents favour private rights and do not recognise farmers' rights at all. The second point is that patenting restricts use of the patented resources. This means that GM technology might be completely harnessed by private companies and developed on the basis of their commercial interests, rather than for the wider public good. It is critical to maintain access to both traditional and new technologies and products, so they can be used to promote benefits for small farmers, poor consumers, and for the general public. A third issue is the potential price implications of granting monopoly intellectual property rights, where the technology and its products can be used by one company only for the period of patent protection. This situation is already affecting people's access to essential drugs in developing countries, where cheaper local production of essential medicines is not allowed by the patent-holding company.

KN: A plant may have 30,000 genes which have been developed from nature and selected by farmers over 10,000 years of agriculture. But if a company puts in one new gene, this will enable the company to own the entire plant variety. Once the company alters a variety and thus develops a 'new' one, they apply for a patent to give them intellectual property rights over a certain sequence of genes. However, patenting rules that were developed for other types of products and technologies are inappropriate for plants, which by their nature reproduce themselves. A CD or a book cannot copy itself, but a rice seed grows into a plant which produces seed in its turn. The farmer using a patented rice variety may have to pay royalties of some sort to the company, even though she or he has grown the plant through investing labour and inputs — but that doesn't count. What counts is that bit of engineering which took place in a laboratory at some point in the past.

Another question is who is holding the patent, and where it is held. For example, a group of US researchers secured a patent on a whole range of varieties of the Andean crop quinoa. The patent was challenged on a number of grounds, in our view quite rightly, but it was granted. As a consequence, farmers who produce quinoa with some of the characteristics that the researchers patented in the USA can't actually export foods from the varieties concerned to the United States, so the patent has had a direct impact on what farmers can grow for export. Quinoa crops are the end result of 10,000 years of farmers' work and knowledge.

PF: Technologies are being developed in a way so as to enforce intellectual property rights with or without patents. It is well-known that a so-called 'terminator gene' has been developed, which has the ability to prevent a plant from reproducing. It hasn't actually been commercially applied to date, and Monsanto recently promised not to release it commercially, but application of this technology in the future would prevent farmers from saving seeds and oblige them to go back and buy seeds from the company each year.

KN: Once that trait has been bred into a number of varieties of a certain plant, companies who own a whole range of other varieties or have breeder's rights over them could breed that trait into all the existing varieties — so even if you don't hold the patent for the 'terminator gene', you can use it to force farmers to buy your seeds.

What do you think of the argument that if GM crops harm nature, they harm women?
KN: That view comes from eco-feminist perspectives, which go too far in conflating women's interests with the interests of 'nature', meaning bio-diversity and eco-systems, especially local eco-systems. If these are being harmed, then poorer farmers, including women, who depend on local resources will be harmed, because the natural ability of eco-systems to cope with pests and diseases will be reduced.

Are there particular problems for women?
KN: Many poor women are involved in farming, particularly in Africa, and the division of labour in agricultural societies means that women have different roles in farming than men. Men often are often chiefly responsible for cash crops, and women for food crops. In Europe, the US, India, Brazil, and other countries, women have a particular stake in these debates because of their role as carers for family members who will consume these crops. The state is making the assumption in many countries that women will provide health services when it fails to.

However, the current technological developments are all investments by the companies in crops that are commercially interesting to farmers. So it is not the women's vegetable patch which is going to be the target of the these big companies; it is going to be either cash crops, or staples which are also cash crops for large farmers. Rural food security in many developing countries may thus be unaffected by the companies' research and development in any direct way. Women are often not making decisions on cash crops, or taking the income from them, but they may be involved in cultivating them. Because they do not necessarily get the benefits from increased production or income, and there is a shift towards more cash crops as a result of further liberalisation and globalisation of markets, their families food security may not improve at all, or even worsen.

Furthermore, the fields with vegetables and household staples that women do grow for family consumption are very often dependent on local biodiversity for natural predators of crop pests, which may well be harmed by an unregulated and badly researched introduction of pesticide-producing GM crops in neighbouring fields. Nevertheless, GM technology has potential for local food security and women's benefit. The nutritional value of staple crops can be enhanced, and research in the non-business sector has already developed a vitamin-A enriched rice strain that could be cross-bred with many other rice varieties, local varieties included. Public-sector research and development could explore possibilities for developing crop varieties which hold nitrogen and thus require less fertiliser, which improve nutritional value, or which improve pest resistance of common homestead-grown vegetables. The latter has the potential effect of less work and more success of women's farming efforts.

PF: Again, it depends on what crops are developed, and how the technology is used. Development of GM technology cannot be left solely in the control of the biotechnology companies. Also, GM technologies can't be seen as the only solution to the problems of poverty; other approaches to agricultural development and food security need to be promoted, such as the sustainable agricultural projects supported by Oxfam which we have already seen achieve some success. Power relations in markets and structural constraints, such as the need for land reform, also need to be addressed.

Protests about the potential imposition of GM crops on producers and consumers have occurred across the world. How have women's organisations and feminist action been part of this?
KN: If you look at India, the feminist movement is at the forefront of protests about GM technologies and patenting, and feminist protests are being organised in a number of countries, including India, Australia, and the UK. Vandana Shiva is particularly well-known in this movement — interestingly, I once met her and she said she considers herself an environmentalist.

It is important that both feminists and environmentalists are in the forefront of the discussions about bio-technology, and that there is united action by activists all over the world on the issue of eco-rights for women and men. In Brazil, in the UK, and in various parts of continental Europe, environmental organisations are leading the protest; these are very often male-dominated, and not terribly strong on their analysis of gender issues.

What should development and relief organisations such as Oxfam be doing to support poor women and men to secure their rights to food and a livelihood?
KN: There are three things that development organisations can do about this. Those with an international voice can try to influence the international policies that actually define the rights of people: policies regulating world trade and national patent systems, but also research policies of governments and multilateral organisations. Second, all kinds of development organisations — both bilateral and non-government — can continue to develop alternatives. Sustainable agriculture has been developed as a response to the first green revolution, and can take on this new bio-technological revolution.

This is not a dream or a theory; it is already happening: community development techniques promote the participation of women and men, and community-based organisations, in the development and use of technologies. This approach is about decentralising research and decision-making on new technologies and their use. The third option is to support people to give voice to their concerns. Women's organisations and consumer movements are already linking in many countries in North and South, and this activity should be supported by donors.

What needs to be done about world trade, to assert the rights of Southern producers to decide what they grow and eat?
PF: The key point is how international trade policy is set at the World Trade Organisation. WTO decisions are currently driven by narrow commercial interests and need to be made consistent with human development policies and strategies. Governments must recognise that trade liberalisation and economic growth are means to an end, not an end in themselves, and that the overall objective should be human development. Oxfam is undertaking advocacy work on international trade and poverty issues and lobbying the

UK government, and the EU, in relation to their role in the WTO. There has been a tendency to suggest that countries can trade their way to food security and that it doesn't matter whether a country produces staple food domestically or imports it. In the case of many poor countries, however, whose foreign exchange reserves may be low or unstable, it is unwise to assume that they will always be able to access food supplies at affordable prices from the international market, to ensure local food security needs. For these countries, where agriculture is an important source of people's livelihoods —especially women's — it is important to protect and promote domestic food production. Development organisations and women's organisations are trying to raise questions about the role of international trade in promoting food security. There will be opportunities to raise these issues at the WTO Ministerial Conference, to be held in Seattle, USA, at the end of this year. Oxfam will be working with other organisations to promote the message that human development objectives should be at the heart of international trade policies.

KN: If a country is deeply indebted and its national harvest fails, it needs to buy food. People are already hungry, so the government has no choice. Nature can jeopardise an entire annual harvest, and the whole situation of a national economy. But markets can also jeopardise the harvest if farmers cannot afford inputs. That fragility of an agricultural economy needs to be considered, and trade policies must be developed which protect national food security. Countries should be granted the right to pursue national food security through a level of protection of their own producers against cheap imports, which is an argument against purely liberalised world trade.

Penny Fowler and Koos Neefjes are Policy Advisers for Oxfam GB on Trade, and Environment and Development, respectively. Contact details: Oxfam GB, 274 Banbury Road, Oxford OX2 7DZ. E-mail pfowler@oxfam.org.uk and kneefjes@oxfam.org.uk

Resources

compiled by Erin Murphy Graham

Books and papers

Gender and Land Use: Diversity in Environmental Practices, Mirjam de Bruijin, Ineke van Halsema, and Heleen van den Hombergh (eds.), Thela Publishers, 1997.
Prinseneiland 305, 1013 LP Amsterdam, The Netherlands.
This book confronts different theoretical approaches which link gender, land use, and the management of the environment from an empirical perspective. Eleven articles discuss case studies from different countries in the southern hemisphere.

A Field of One's Own: Gender and Land Rights in South Asia, Bina Agarwal, Cambridge University Press, 1994.
The Pitt Building, Trumpington Street, Cambridge CB2 1RP, UK / 40 West 20th Street, New York, NY 10011-4211, USA
Bina Agarwal's comprehensive and rigorous study of women's land rights in South Asia argues in favour of women's ownership and control of land and property on grounds of gender equity, women's need for empowerment, economic efficiency, and family welfare. It also focuses on women's activism and resistance to their marginalisation from land rights, it is aimed at a varied audience of scholars, students, policy-makers and activists.

Agriculture, Women, and Land: The African Experience, Jean Davison (ed.), Westview Press, 1988.
5500 Central Avenue, Boulder, CO 80310, USA.
This collection of 12 articles on topics relating to women and agriculture in Africa emphasises the diversity and complexity of social relations which shape women's access to agricultural land.

Women, Land, and Authority: Perspectives from South Africa, Shamim Meer (ed.), Oxfam, 1997.
BEBC, PO Box 1496, Parkstone, Dorset, BH12 3YD, UK.
The question of land lies at the heart of South Africa's democratic transition. This book brings together research by the National Land Committee of South Africa to explore women's attitudes to land, and conditions of subsistence, labour, and housing.

Women, the Environment and Sustainable Development: Towards a Theoretical Synthesis, R Braidotti et al., Zed Books.
7 Cynthia Street, London N1 9JF, UK/165 First Avenue, Atlantic Highlands, NJ 07716, USA.
This book examines alternative visions of development, including Women, Environment and Development (WED) and Eco-feminism, aiming to disentangle the various positions put forward by major actors and

to clarify the political and theoretical issues at stake in the debates on women, the environment, and sustainable development.

Ecofeminism, Maria Mies and Vandana Shiva, Zed Books, 1993.
This pathbreaking book challenges current models of development, providing an overview of eco-feminist ideas from a Southern perspective.

Gender, Household Food Security and Coping Strategies, Julie-Koch Laier et al., Institute of Development Studies, 1996.
IDS, University of Sussex, Brighton, Sussex BN1 9RE, UK.
Recognising that women play a crucial yet constrained and undervalued role in ensuring food security, this annotated bibliography draws together the literature on household food security in Sub-Saharan Africa and South Asia; it contains a special section on agricultural production.

Structural Adjustment and African Women Farmers, Christina H Gladwin, University of Florida Press, 1991.
15 Northwest 15th Street, Gainesville, Florida 32603, USA.
Analyses the impact of structural adjustment programmes on women farmers in several African countries, presenting evidence from noted social scientists who take positions on both sides of the debate.

Women Wielding the Hoe: Lessons from Rural Africa for Feminist Theory and Development Practice, Deborah Fahy Bryceson (ed.), Berg Publishers, 1995.
150 Cowley Road, Oxford OX4 1JJ, UK.
This volume's 12 chapters examine themes such as agricultural production, reproduction, women's workload in their multiple roles, and women and AIDS.

Women Plantation Workers: International Experiences, Shobhita Jain and Rhoda Reddock (eds.), Berg Publishers, 1998.
Using case studies, this collection of essays describes and analyses the experiences of women workers and the socio-economic systems of plantations world-wide.

Gendered Fields: Rural Women, Agriculture, and Environment, Carolyn E Sachs, Westview Press, Inc., 1996.
This highly theoretical volume explores several themes, including feminist theory and rural women, rural women and nature, connections to the land, work with plants and animals, and women on family farms.

Women and Sustainable Development in Africa, Valentine Udoh James, Praeger Publishers, 1995.
88 Post Road West, Westport, Connecticut 06881, USA.
Bringing together a number of scholars to articulate the significance of women's contribution to Africa's development, this volume presents ten chapters that explore themes relating to women and development in Africa, several of which focus on agriculture.

Women in Agriculture: What Development Can Do, Mayra Buvnic and Rekha Mehra, International Center for Research on Women, 1990.
1717 Massachusetts Avenue, NW, Suite 302, Washington, DC 20036, USA.
This study reviews the research evidence on women's roles in farming, the impact of technology on women farmers, and development projects for rural women.

Women and Food Security: The Experience of the SADCC Countries, Marilyn Carr (ed.), IT Publications, 1991.
Authors consider women's access to, and use of, improved food technologies, and relates these to the external contexts of development policy, markets, and infrastructure. Focuses on southern Africa, but has resonance beyond this region.

Women and Water-Pumps in Bangladesh: The Impact of Participation in Irrigation Groups on

Women's Status, Barbara van Koppen and Simeen Mahmud, IT Publications, 1996.
A study of women's participation in irrigation projects, drawing on the experience of 35 female and mixed-sex irrigation groups supported by six NGOs. Provides policy recommendations.

Women and the Transport of Water, V Curtis, IT Publications, 1986.
This paper looks at this most time-consuming task of rural women, examining the overall scale of the problem and looking at Kenya in particular.

Tools for the Field: Methodologies Handbook for Gender Analysis in Agriculture, Hilary Sims Feldstein and Janice Jiggins (eds.), Kumarian Press, 1994.
630 Oakwood Avenue, Suite 119, West Hartford, CT 06110-1592, USA.
This handbook offers a practical set of tools for individuals working on gender analysis in agriculture, with 39 case studies covering Latin America, Asia, and Africa.

Women in Agriculture: A Guide to Research, Marie Manman and Thelma H Tate, Garland Publishers, 1996.
Taylor and Francis, 47 Runway Road, Suite G, Levittown, PA 19057, USA.
This annotated bibliography contains over 700 resources, including books, journal articles and titles, dissertations, and electronic indexes and resources. Topics include the sexual division of labour in agriculture, decision-making on the farm, women's role in policy implementation, and the education of women in agriculture.

Women in Agriculture: Gender Issues in South Asian Farming, available from SAARC Agricultural Information Centre, 1993.
BARC Complex, Farmgate, Dhaka, Bangladesh.
This annotated bibliography contains over 800 citations in 30 categories, which include co-operatives, fisheries and aquaculture, irrigation, fertilising, and agricultural economics and policies.

Women in Agriculture: Farming for Our Future, 1994 International Post-Conference Proceedings.
Available from the University of Melbourne, International Conference Committee, RMB 7395, Sale, Victoria 3850, Australia.
Themes explored include women in agriculture, production and the environment, sustainable development, and computers for farm management; conference outcomes and recommendations are also included.

Women in the Third World: An Encyclopaedia of Contemporary Issues, Nelly P Stromquist (ed.), Garland Publishing, 1999.
This concise reference work was written by more than 80 international experts. Two sections focus on women and the environment and women and production, containing several articles on women, land, and agriculture, including women in agricultural systems and women's roles in natural resource management.

Organisations

Food and Agriculture Organisation of the United Nations (FAO), Viale delle Terme di Caracalla, 00100 Rome, Italy. Tel: +39 06 57051, Fax: +39 06 5705 3152
The FAO is the largest autonomous UN agency that aims to promote agricultural development and food security.

Women and Rural Economic Development (WRED), The Learning Centre, 423 Erie St, Stratford, Ontario N5A 2N3, Canada.
Tel: +1 519 273 5017, Fax: +1 519 273 4826
http://www.wred.org
WRED is a Canadian NGO that aims to support the sustainability of rural Ontario communities by promoting economic opportunity and programmes that enhance business development, life skills, networking,

access to capital, information, and markets, business diversification, and awareness of rural community economic development. It also contains an online bibliography.

Global Network for Rural Women, PO Box 1634 M, Melbourne 3001, Australia.
Tel: +61 (0)3 5634 2634
The Rural Women's Network links rural women from around the world. It publishes the newsletter 'Global News', featuring articles with an international perspective.

International Federation of Agricultural Producers (IFAP-FIPA), 60 rue St. Lazare, 75009 Paris, France. Tel +33 1 4526 0553, Fax: +33 1 4874 7212, E-mail: info@ifap.org
IFAP is the official representative of the world's farmers recognised by the UN. Its aim is to allow farmers to be heard and to influence decision-making by international organisations such as the FAO, the Organisation for Economic Co-operation and Development (OCED), the World Trade Organisation (WTO), and the World Bank.

Associated Country Women of the World, Vincent House, Vincent Square, London SW1 2NB, England. Tel: +44 (0)171 8348 635, Fax: +44 (0)171 2336 205.
An organisation representing more than 70 countries around the world, Associated Country Women of the World is active in rural areas and attempts to work with women in rural areas throughout the world to overcome the problems facing them.

International Research and Training Institute for the Advancement of Women (INSTRAW), Avenida César Nicolás Penson, 102-A, PO Box 21747, Santo Domingo, Dominican Republic. Tel: +1 809 685 211.
INSTRAW's primary objective is to ensure that sustained attention is given to the integration of women in development at all levels, including in agriculture. Among other topics, INSTRAW focuses on food, natural resources, and agribusiness.

The Association for Women in Development (AWID), 666 11th Street, NW, Suite 450, Washington, DC 20001, USA. Tel +1 202 628 0440, Fax: +1 202 628, E-mail: awid@awid.org
AWID is an international membership organisation committed to gender equality and a just and sustainable development process. Through AWID membership, scholars, practitioners, and policy-makers from around the world discuss and share ideas concerning development strategies and programmes.

Web resources

Global Network for Rural Women
http://home.mira.net/~faawagri/global/index.html
The Global Network for Rural Women was established by agricultural and rural women from around the world. Through online communication, printed newsletters, personal mail, and international meetings, they have ready access to a global forum to debate vital issues such as trade agreements, food security, and women's rights.

International Federation of Agricultural Producers: Women in Agriculture
http://www.ifap.org/women.html
Founded in 1946, the International Federation of Agricultural Producers is the international farmers' organisation. The IFAP Standing Committee on Women in Agriculture is its main decision-making forum of rural women. This site also contains useful information on the World Rural Women's Day.

Women and Population (FAO)
http://www.fao.org/WAICENT/FAOINFO/SUSTDEV/WPdirect/default.htm
This FAO site includes online analysis papers on rural gender issues, a bibliography of the FAO's Plan of Action for Women in Development, as well as current activities on gender and agriculture.

Women in AG
http://www.agriculture.com/sfonline/archive/sf/women/wagcont.html
Mainly of interest to rural and farm women in the USA, this web site contains articles, links, and a discussion group for women in agriculture.

Women in Agriculture and Rural Life: An International Bibliography
http://www.nal.usda.gov/afsic/wia/women.htm
An online bibliography published to coincide with the Second International Conference on Women in Agriculture, held in July 1998. It is divided into three sections: Women on the Land, Women as Agricultural Professionals, and Bibliographies and Non-Media Print. Contains links to the Alternative Farming Systems Information Center and The Economic Research Service of the US Department of Agriculture.

The World Agricultural Information Centre (WAICENT)
http://www.fao.org.
Thhe FAO's internet programme on information management and dissemination, WAICENT provides access to FAO's data and specialised information on topics of global relevance, including gender and sustainable development.

Women in Agriculture
http://www.wia.usda.gov/index.htm
The Women in Agriculture (WIA) website was created to follow-up on the First and Second International Conferences for Women in Agriculture (ICWA), and to prepare for the Third International Conference which will be held in Spain in 2002. Serves as a communications network for women in agriculture where they can share information, trade ideas, and increase their opportunities and resources. Contains translations in French, German, Italian, Spanish, and Portuguese.

E-mail lists

AGWOMEN-L is an Australian list for women involved in agriculture and those interested in the issues facing women in rural Australia. To subscribe, send an e-mail stating 'Subscribe agwomen-1' to majordomo@peg.apc.org

Development-Gender, a moderated list run by the Gender, Research, and Training unit of the School of Development Studies at the University of East Anglia, UK, discusses many issues relevant to gender, agriculture, and rural development. To subscribe, send an e-mail stating 'Subscribe development-gender' to mailbase@mailbase.ac.uk

Rural Women, hosted by the Global Network for Rural Women, provides a discussion list for rural women in order to debate issues such as trade agreements, food security, and women's rights, as well as the opportunity to share cultural, social, and practical agricultural information. To subscribe, send an e-mail stating 'subscribe ruralwomen' to majordomo@lists.vicnet.au

Women in Agriculture is an e-mail list service with an online forum for discussion, exploration, and support. It is an excellent resource for women world-wide who are involved in agriculture. Subscribers to the list service will receive information, notices, and updates regularly by e-mail. To subscribe, send an e-mail that includes your e-mail address to: wialist@rus.usda.gov

Index to Volume 7

Adamu, Fatima L, A double-edged sword: Challenging women's oppression within Muslim society in Northern Nigeria, 6:1, 56

Ahmed, Sadia, Islam and development: Opportunities and constraints for Somali women, 6:1, 69

Arun, Shoba, Does land ownership make a difference? Women's roles in agriculture in Kerala, India, 7:3, 19

Burlet, Stacey, Gender relations, 'Hindu' nationalism, and NGO responses in India, 6:1, 40

Chapman, Katie and Gill Gordon, Reproductive health technologies and gender: Is participation the key?, 6:2, 34

Chilimampunga, Charles, The denigration of women in Malawian radio commercials, 6:2, 71

Clancy, Kathleen and Sharon Harper, 'The way to do is to be': Exploring the interface between values and research, 6:1, 73

Costa, Ana Alice, Elizete Passos, and Cecilia Sardenberg, Rural development in Brazil: Are we practising feminism or gender? 7:3, 28

Dolan, Catherine S, Conflict and compliance: Christianity and the occult in horticultural exporting, 6:1, 23

Fawcett, Ben and Shibesh Chandra Regmi, Integrating gender needs into drinking-water projects in Nepal, 7:3, 62

Fonchingong, Charles, Structural adjustment, women, and agriculture in Cameroon, 7:3, 73

Foster, Maggie, Supporting the invisible technologists: The Intermediate Technology Development Group, 6:2, 17

Fowler, Penny, Interview with Koos Neefjes of Oxfam GB, 7:3, 80

Gajjala, Radhika and Annapurna Mamidipudi, Cyberfeminism, technology, and international 'development', 6:2, 8

Gordon, Gill and Katie Chapman, Reproductive health technologies and gender: Is participation the key?, 6:2, 34

Harper, Sharon and Kathleen Clancy, 'The way to do is to be': Exploring the interface between values and research, 6:1, 73

Hashim, Iman, Reconciling Islam and feminism, 7:1, 7

Humphreys, Rachel, Skilled [workers] or cheap labour? Craft-base[d production] as an alternative to female urban migration in northern Thailand, 6:2, 56

Izumi, Kaori, Liberalisation, gender, and the land question in sub-Saharan Africa, 7:3, 9

Macey, Marie, Religion, male violence, and the control of women: Pakistani Muslim men in Bradford, UK, 6:1, 48

Mamidipudi, Annapurna and Radhika Gajjala, Cyberfeminism, technology, and international 'development', 6:2, 8

Naylor, Rachel, Women farmers and economic change in northern Ghana, 7:3, 39

Neefjes, Koos, Interview with Penny Fowler of Oxfam GB, 7:3, 80

Otsyina, Joyce A and Diana Rosenberg, Rural development and women: What are the best approaches to communicating information? 6:2, 45

Passos, Elizete, Ana Alice Costa, and Cecilia Sardenberg, Rural development in Brazil: Are we practising feminism or gender? 7:3, 28

Prabhu, Maya, Marketing treadle pumps to women farmers in India, 6:2, 25

[Regmi, Shibesh] Chandra and Ben Fawcett, [Integrating gen]der needs into drinking-water projects in Nepal, 7:3, 62

Rosenberg, Diana and Joyce A Otsyina, Rural development and women: What are the best approaches to communicating information?, 6:2, 45

Sardenberg, Cecilia, Ana Alice Costa, and Elizete Passos, Rural development in Brazil: Are we practising feminism or gender? 7:3, 28

Saul, Rebecca, No time to worship the serpent deities: Women, economic change and religion in north-western Nepal, 6:1, 31

Schreiner, Heather, Rural women, development, and telecommunications: A pilot programme in South Africa, 6:2, 64

Tripp, Linda, Gender and development from a Christian perspective: Experience from World Vision, 6:1, 62

Walker, Bridget, Christianity, development, and women's liberation, 6:1, 15

Whitehead, Ann, 'Lazy men', time-use, and rural development in Zambia, 7:3, 49